MAKERS OF Science

VOLUME 1

Michael Allaby & Derek Gjertsen

OXFORD
UNIVERSITY PRESS

Consulting editors

Dr. John Gribbin
SUSSEX UNIVERSITY, U.K.

Willem Hackmann
MUSEUM OF THE HISTORY OF SCIENCE, OXFORD, U.K.

Published in the United States of America by
Oxford University Press, Inc.
198 Madison Avenue
New York, NY 10016
www.oup.com

Planned and produced by
Andromeda Oxford Limited
11-13 The Vineyard
Abingdon
Oxon OX14 3PX
UK
www.andromeda.co.uk

Library of Congress Cataloging-in-Publication Data

Allaby, Michael.
Makers of Science / Michael Allaby & Derek Gjertsen.
Includes bibliographical references and indexes.
ISBN 0-19-521680-6 (set)
1. Scientists--Biography. 2. Science--History. I. Gjertsen, Derek. II. Title

Q141 .A44 2002
509.2'2--dc21
[B] *20011048396*

Project Directors *Susan Kennedy, Peter Lewis*
Managing Editor *Penelope Isaac*
Design *Steve McCurdy, Chris Munday*
Editors *John O. E. Clarke, Celia Coyne*
Picture Manager *Claire Turner*
Picture Researcher *Liz Eddison*
Production *Clive Sparling*
Index *Ann Barrett*

Printed in Singapore

Contents

Previous page: An optical photograph shows the giant spiral galaxy of Andromeda, known to Western astronomers since the 17th century.

Right: A model by Johannes Kepler shows how he imagined that the six planets then known, and the five regular solids described by the Greek mathematician Euclid (active c. 300 BC), might be arranged to form a well-ordered planetary system.

β
α

Using this set

This fully illustrated five-volume set provides a fascinating overview of Western science and scientific ideas from the 4th century BC to the present day. The lives of more than 40 key scientists—men and women whose discoveries were crucial to the development of science—are charted chronologically in chapters. The **main text** discusses their life and work in full, and a **key dates** panel gives an at-a-glance summary of their careers. Contemporary quotations and striking paintings and photographs create a vivid picture of the subjects and the world in which they worked. Lively **feature boxes** look at a wide variety of topics ranging from detailed aspects of scientists' work to the broader events that influenced or shaped their lives—from important scientific institutions to advances in scientific instruments and apparatus. **Biographical boxes** within each chapter discuss the careers of significant contemporaries. Where appropriate, **color diagrams** are used to illustrate key scientific experiments or theories. At the end of each chapter two **timelines** show the scientific and the political and cultural events of the period. An **index** completes each book.

The fifth volume contains further reference resources in the form of nearly 300 additional brief biographies describing the men and women who have shaped our scientific world, a glossary, a list of further reading and reliable websites, and a comprehensive set index.

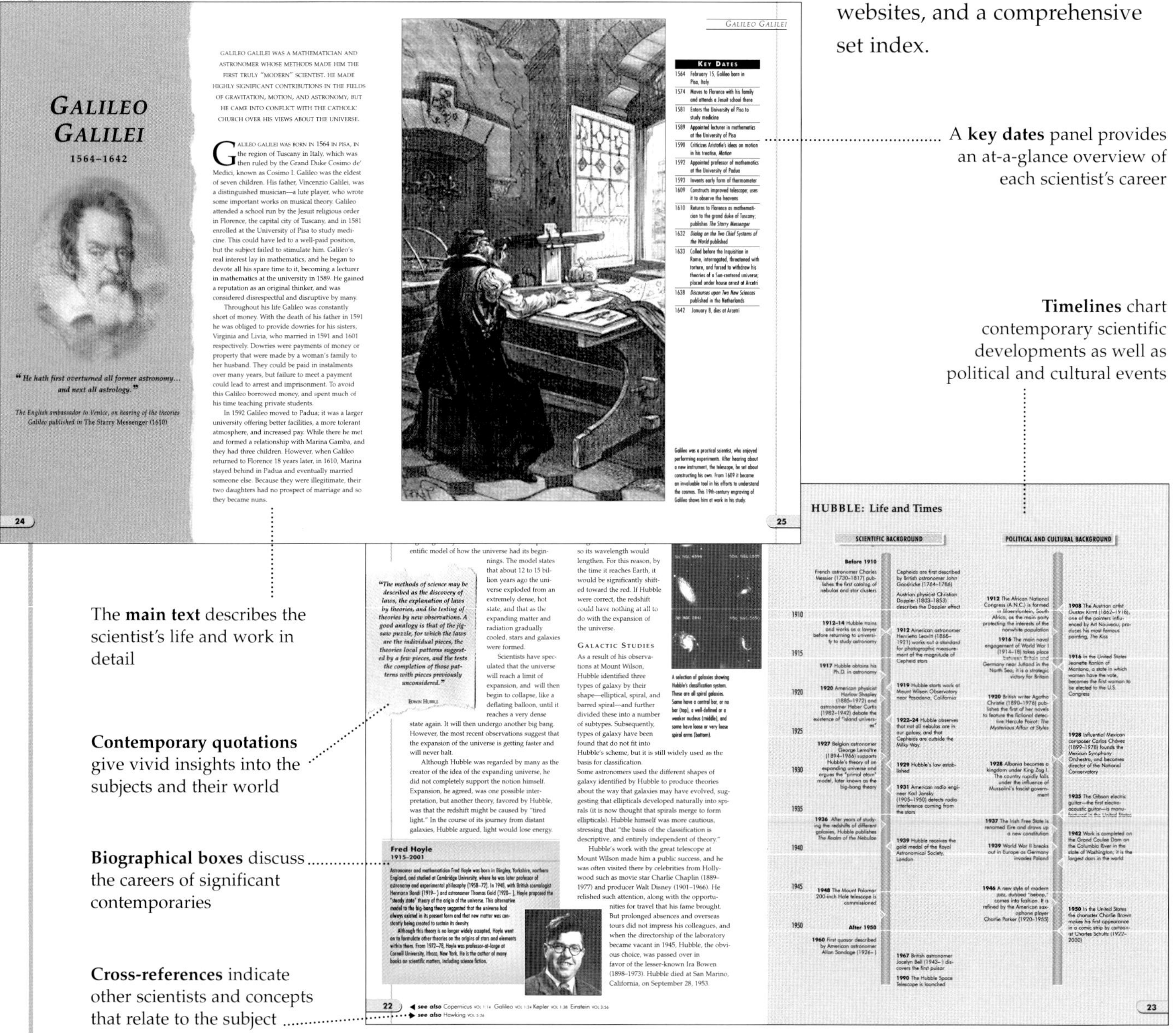

A **key dates** panel provides an at-a-glance overview of each scientist's career

Timelines chart contemporary scientific developments as well as political and cultural events

The **main text** describes the scientist's life and work in detail

Contemporary quotations give vivid insights into the subjects and their world

Biographical boxes discuss the careers of significant contemporaries

Cross-references indicate other scientists and concepts that relate to the subject

About this volume

It has taken more than 2,000 years to build up the knowledge of science that we have today. This volume takes us from the world of the ancient Greeks in the 4th century BC to 18th-century France. It begins with the Greek thinker Aristotle, who based his ideas about the natural world (such as the behavior of animals, plants, and the weather) only on what he could observe with his own eyes—the origins of scientific method. More than a thousand years later, as the Middle Ages were coming to a close, the astronomers Copernicus, Galileo, and Kepler observed that the Earth and planets circle around the Sun, and not the other way around. This directly contradicted what the Christian Church, the dominant body of thought in medieval Europe, taught about the universe; the astronomers' findings—breaking with the tyranny of the past—brought science forward into the age of reason. With his mathematical work on light, motion, and gravity, Newton went still further in increasing scientific understanding of how the universe works. Today these great topics all come under the heading of physics; Newtonian physics was to remain unchallenged until the beginning of the 20th century and is still fundamental to scientific understanding of the physical world. Harvey, who studied the circulation of the blood, and Linnaeus, who studied and classified the plant kingdom, gave new insights into the science of living organisms (physiology and botany respectively). Finally, Lavoisier discovered that water is made up of two separate elements, oxygen and hydrogen. His work laid the foundations of modern chemistry.

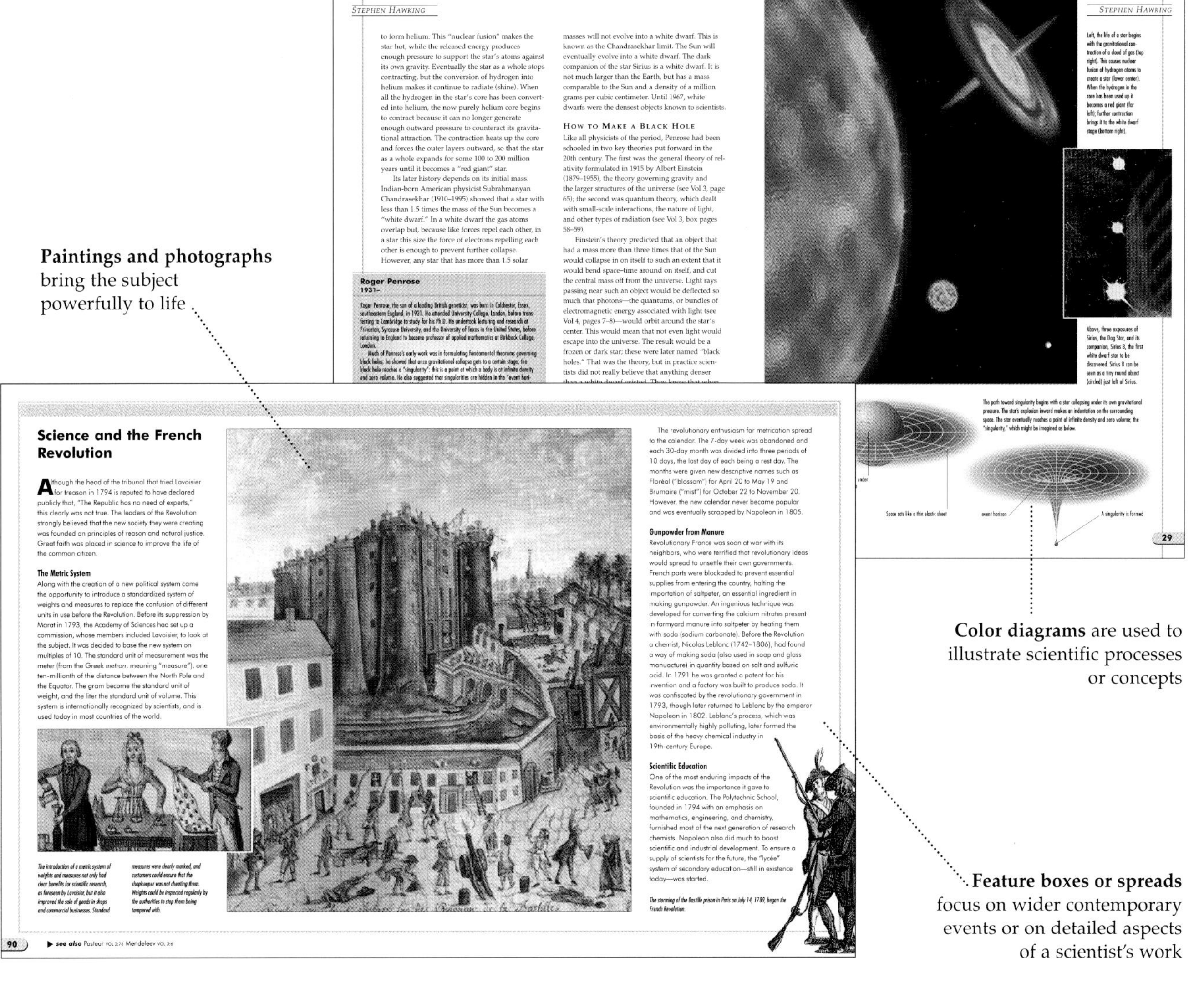

Paintings and photographs bring the subject powerfully to life

Color diagrams are used to illustrate scientific processes or concepts

Feature boxes or spreads focus on wider contemporary events or on detailed aspects of a scientist's work

Aristotle

384 BC–322 BC

"...*More trust should be put in the evidence of observation rather than in theories, and in theories only insofar as they are confirmed by the observed facts.*"

ARISTOTLE
On the Generation of Animals
(4TH CENTURY BC)

THE ANCIENT GREEK PHILOSOPHER ARISTOTLE POSSESSED AN OUTSTANDING INTELLECT. HIS WRITINGS—ON SUBJECTS FROM POLITICS TO ZOOLOGY—SPANNED ALL BRANCHES OF HUMAN KNOWLEDGE. ALTHOUGH HIS IDEAS HAVE NOT ALWAYS STOOD THE TEST OF TIME, ARISTOTLE'S APPROACH HAD A PROFOUND INFLUENCE ON LATER WESTERN AND ISLAMIC THOUGHT AND SCIENCE.

ARISTOTLE WAS BORN IN 384 BC IN STAGIRA, a town on the Chalcidice peninsula in northern Greece. His father was Nichomachus, a physician at the nearby Macedonian court of King Amyntas III (d. 370/369 BC). Macedonia was about to become the greatest power in Greece. In 349 BC Amyntas's son Philip II (382–336 BC) took control of Chalcidice, and by 339 BC Macedonia dominated the whole of Greece, including Athens. Once the most important city in Greece, Athenian power had begun to decline about 30 years before, but it was still the leading center of learning in the Greek world.

Early Life

In 367 BC, after his father's death, Aristotle went to study at the Academy in Athens. This famous school had been founded 20 years earlier by the philosopher Plato (c. 428–c. 348 BC). Aristotle stayed there for the next 20 years, until Plato's death in 348 BC. Plato had decreed that his nephew, Speusippus (c. 407–339 BC), should head the Academy, and at this time Aristotle left Athens. It is unclear whether this was because he was upset at being passed over or because feelings in Athens against Macedonians were running high at the time. He was to spend the next 12 years away.

From Athens, Aristotle traveled to Assos, a town opposite the north shore of the island of Lesbos, on the coast of what is now northwest Turkey. Here Aristotle befriended another philosopher, Theophrastus (c. 372–c. 287 BC), and together they moved to Mytilene, the chief city on Lesbos. Aristotle studied the natural history there, spending many hours at the inland lagoon of Pyrrha in the middle of the island.

Schools of Learning in Athens

Permanent schools were still a fairly recent development in Greece when Aristotle enrolled at Plato's Academy in 367 BC. Before this, individual teachers known as Sophists (from the Greek word meaning "wise" or "expert") would travel throughout the Greek world, settling for short periods in a particular spot to take in a few pupils for a fee.

Like other schools in Athens, Plato's Academy was an extension of a public gymnasium: exercise was held to be a key part of the education of young Athenian males (girls were excluded, not being considered fit for schooling). It was a center for philosophical, mathematical, and scientific study. Groves of olive trees served as classrooms and students took part in philosophical discussions.

Plato, perhaps the most influential philosopher who has ever lived, set up his Academy in 387 BC on his return to Athens after a long period abroad following Socrates's trial and death. Socrates (469–399 BC) had been Plato's teacher. The most famous philosopher of his day, Socrates left no writings of his own. His ideas have come down to us in the form of dialogs (conversational exchanges) written down by Plato, through which Plato also expounded his own philosophical theories. Socrates's ideas were troubling to many in Athens. In 399 BC he was charged with "neglect of the gods" and "corrupting the youth" of Athens, found guilty, and condemned to death by drinking hemlock, a deadly poison.

In this Roman mosaic found at Pompeii, Plato is depicted teaching at his Academy. Using a stick to draw figures in the sand, he instructs fellow philosophers in geometry.

In 342 BC Aristotle returned to Macedonia, having been appointed to teach Philip II's 14-year-old son, Alexander. After his father's death, Aristotle's royal pupil would conquer most of the known world and go down in history as Alexander the Great (356–323 BC). In about 339 BC Aristotle left the Macedonian court and returned to Stagira, his birthplace, where he continued to record as much as he could about the natural world.

The Greek gymnasiums were schools where male athletes over the age of 18 received training. The gymnasiums provided an all-around education for young men, teaching philosophy, music, and literature as well as athletics. Students regularly took part in public sports festivals, competing for trophies such as this ceramic bowl with a sporting theme.

KEY DATES

384 BC	Born in northern Greek colony of Stagira, the son of a physician
367 BC	Moves to Athens to study at Plato's Academy
345–42 BC	Devotes time to study of natural history on Lesbos
342 BC	Appointed tutor to Alexander the Great in Macedonia
335 BC	Returns to Athens where he establishes the Lyceum
323 BC	Leaves Athens, succeeded by Theophrastus as head of Lyceum
322 BC	Dies on island of Euboea

School of Athens (c. 1510–12) by the Italian painter Raphael (1483–1520). At the center, Plato points to the heavens and Aristotle to Earth as they discuss where the eternal truths lie. The period of the Renaissance, when this picture was painted, saw a great revival of interest in the philosophical ideas of classical Greece.

The Final Years

Aged about 49, Aristotle returned to Athens in 335 BC and, possibly financed by Alexander the Great, founded his own school, called the Lyceum. Aristotle's followers became known as peripatetics, from the Greek word *peripatetikos* (to walk to and fro), because of their habit of shadowing his footsteps as he paced up and down while lecturing.

After the death of Alexander in 323 BC there was another outbreak of anti-Macedonian feeling in Athens. Aristotle seems to have felt threatened by this. He is supposed to have said that he could not allow the Athenians "to sin twice against philosophy," a reference to the execution of the philosopher Socrates in 399 BC, and he retired to the isle of Euboea. Aristotle died there soon afterward, in 322 BC, and his friend Theophrastus succeeded him as head of the Lyceum in Athens.

Science before Aristotle

The ancient Greeks were highly artistic, literate, and educated people, and science and philosophy had been studied from long before Aristotle's time. Greek thinkers were especially interested in finding rational explanations for natural events such as thunder storms or earthquakes. The founder of Greek science and philosophy was traditionally held to be Thales of Miletus (c. 624–c. 545 BC). He had declared that everything was made of water, probably because he was able to observe that liquid water is transformed into air as steam and into a solid as ice. Other Greek scientists believed that all natural objects are made from four basic elements—earth, air, water, and fire—while yet others argued that the world is composed of indivisible units of matter they called atoms.

Aristotle viewed the universe as a sphere (left), with the Earth at its center, the stars fixed at the edge, and crystal spheres carrying the Sun, the Moon, and the other planets in between. The 15th-century German painting (below) shows Aristotle the cosmologist with his attentive pupils. Cosmology is the study of the origin and nature of the universe.

Greek thought was highly rational—that is, ideas were reached by systematically collecting all the known facts about a subject and placing them within an overall scheme. We call this approach "scientific" (science simply means "knowledge"). The high value given to literacy meant that ideas could be written down and passed around for general criticism. The so-called "Socratic dialogue," in the form of a philosophical discussion among a group of people on a particular topic such as "knowledge," or "truth," or "justice," developed out of this practice.

Aristotle's Writings

Aristotle's writings have had long-lasting influence partly because so many of them remained available, unlike those of his great rival, Plato, which were only rediscovered in the 15th century. Also in contrast to Plato, Aristotle did not write any dialogs. Instead, his ideas exist mostly in the form of lecture notes or texts for students. His surviving writings cover almost every field of knowledge—logic, ethics, biology, zoology, cosmology (the nature of the universe), poetry, physics, and pyschology. Unlike modern scientists, Aristotle and his contemporaries did not use scientific experimentation to confirm their theories about things. But Aristotle always tried to support his ideas with evidence based on his own observations and practical experience. This is called an empirical approach.

Planetary Puzzles

Most people in the ancient world believed the Earth was shaped like a ball, but Aristotle was the first person to prove this scientifically by observing that the shadow of the Earth is round when it appears on the Moon during an eclipse. Only a sphere would do this—if the Earth were a flat disk, for example, it would make an oval shadow.

In common with most thinkers of his day, Aristotle believed that the Earth lay at the center of the universe. He decided that Earth must be fixed, because if it were moving it would be liable to fierce winds and unsteadiness. He also believed that the stars and planets were carried on transparent spheres, like shells around Earth. Each sphere, he thought, must carry a separate heavenly body: the Moon, the Sun, Mercury, Venus, Mars, Jupiter, and Saturn (the only planets known at the time).

According to Aristotle, the spheres rotated daily around a stationary Earth with steady (uniform) circular motion but at different speeds. They were made of a rigid, transparent material. Beyond these an eighth sphere held the fixed stars: Aristotle

Zoological Studies

Aristotle was the first person to try to describe the animals of the known world in a systematic way: he set about studying hundreds of creatures in order to classify and understand them. In works such as *History of Animals* and *The Generation of Animals*, he carefully noted their physical characteristics, but also tried to consider how animals reproduce, how they move, their physical structure (anatomy), their processes and functions (physiology), and their behavior.

Aristotle was a keen observer of nature, but also believed that there was a reason plants and animals were as they were: he thought that they had natural purposes or "ends," and these dictated the forms they develop. This belief is called teleology, from the Greek word *telos*, meaning "end." "Spiny lobsters have tails because they swim about," Aristotle wrote, "and so a tail is of use to them, serving them for propulsion like an oar." This view of nature was extraordinarily influential, right up until the 19th century, when the English naturalist Charles Darwin (1809–1882) showed that the characteristics developed by animals and plants come about not through preordained purpose but through "natural selection." This allows the survival of those animals and plants that are best suited to their surroundings, and the development of new species (see Vol 2, page 62–65). Natural selection is often loosely referred to as "the survival of the fittest."

In his work Aristotle described more than 500 different species. He tried not to accept reports that he could not confirm: he wrote, cautiously, that "if we are to believe Ctesias," [a 5th-century BC Greek historian and physician] "there is such an animal as the manticora, with a triple row of teeth in both upper and lower jaw, as big as a lion, hairy, resembling a man in its face and ears." The animal that Ctesias described so fancifully was probably an Indian tiger.

An 18th-century etching shows Aristotle and a colleague surrounded by an array of exotic animals. Aristotle observes the creatures intently, making notes or sketches.

In many regards Aristotle's biological work was groundbreaking. It includes a description of the joints of an elephant's leg, discrediting the common belief that elephants had to lean against trees to sleep. He also made a detailed study of life within a beehive, which was mostly well observed, though he thought the hive was ruled by a male, not a female, bee.

Aristotle's most precise and detailed work was in his study of marine life. His assertion that the female catfish left her eggs to be cared for by the male was scorned, but it was among several of his claims that 19th-century scientists found to be absolutely accurate.

decided that this must move all the other spheres, and called it the "prime mover." Not all Greek thinkers agreed; for example, Aristarchus of Samos (310–230 BC) believed that the Sun, not the Earth, lay at the center of the universe. But Aristotle's system became generally accepted and was to dominate European thought until the 15th century.

How Things Move

Aristotle also studied motion, beginning with the simple assumption that "Whatever is moved is moved by something." So, to the question, "Why do things move?" Aristotle answered, "Because something moves them." This belief had a great influence on scientists for centuries, until it was

Medieval Islamic scholars continued to study Aristotle's teachings at a time when they had been mostly forgotten in the West. This is a 13th-century Arabic translation of one of Aristotle's zoological works. It was the translation of texts from Arabic into Latin in the 12th century that reintroduced much of Aristotle's scientific work to Western scholars.

disposed of by Galileo Galilei (1564–1642) and Isaac Newton (1642–1727). Even in Aristotle's time it clearly had its limitations, as it could be seen that many things are able to move without a mover—for instance, apples fall to the ground out of trees.

Aristotle dealt with this difficulty by defining three types of motion that take place on Earth. Living creatures move because they choose to ("voluntary" motion). The second kind of motion he described as "natural." Rejecting the ideas of the atomists, he held that objects were composed of the four different elements—earth, water, air, and fire. Earth, the heaviest, lay at the center of the cosmos, surrounded by water, then air, then fire. When moved from their natural place, objects will try to return. This explains why air bubbles rise through water, or earth falls through air.

Aristotle noted that objects speed up, or accelerate, as they fall, but thought this was because the nearer they got to their natural place, the greater the force attracting them. He wrongly believed that heavy objects fall faster than lighter ones, reasoning that the more matter an object contains, the more quickly it returns to its natural place. This view was unchallenged until the 17th century when Galileo proved that a heavy and a light object dropped at the same time hit the ground more or less together (see page 26).

According to Aristotle's theory, an object such as an arrow would fall to the ground as soon as the "mover" (the archer) sets it in motion. As this clearly did not happen, Aristotle concluded that the medium it passed through (in this case, air) must somehow move the object. He called this "violent" motion.

As already noted, Aristotle believed that heavenly bodies such as the stars and planets moved in a circular uniform motion. This form of movement was unlike any of the types of motion that take place on Earth, and from this he reasoned that the stars and planets are not subject to the laws that control the behavior of objects on Earth. Therefore, they could not be composed of the four Earthly elements of earth, water, air, and fire, but of a fifth, entirely different element. Later philosophers came to call this fifth element quintessence (from *quinta*, Latin for fifth, and *essentia*, meaning the being or essence of a thing). Aristotle divided the heavens (what we call space today) into two separate and distinct realms: the "superlunary" (above the Moon) and the

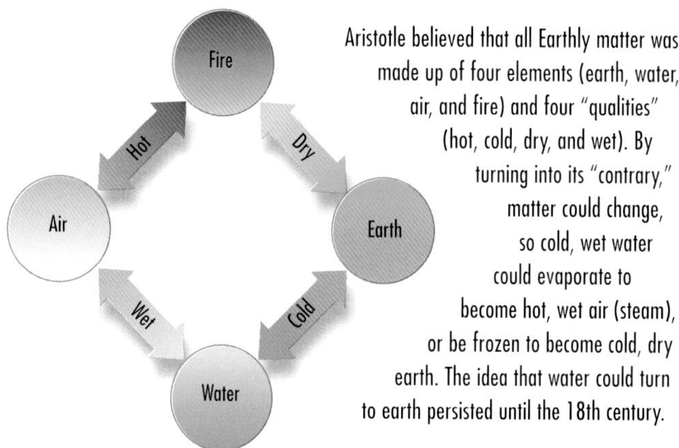

Aristotle believed that all Earthly matter was made up of four elements (earth, water, air, and fire) and four "qualities" (hot, cold, dry, and wet). By turning into its "contrary," matter could change, so cold, wet water could evaporate to become hot, wet air (steam), or be frozen to become cold, dry earth. The idea that water could turn to earth persisted until the 18th century.

"sublunary" (below the Moon). The sublunary realm was subject to the same physical and natural laws that operate on Earth, but the superlunary realm was governed by its own laws. Aristotle further believed that the Earthly elements change their form by losing one of a pair of opposite properties, known as "contraries," and gaining the other (see figure above). Quintessence, however, does not possess these contraries. Because Aristotle believed that change requires contraries, he reasoned that there can be no change without them; therefore, he concluded, the heavens must be unchangeable. From this, he reasoned that any object moving in the skies such as a comet or lightning must occupy the sublunary realm.

ARISTOTLE'S LEGACY

Why should we be interested in Aristotle's ideas about the physical universe today, when we know so many of his theories are wrong? The answer is that he laid the foundations of scientific method through observation and deduction, and determined many of the paths along which the history of science and scientific thinking would develop in the coming centuries. Whether concerned with notions of matter, motion, change, or the cosmos, all scientific discussion for 1,500 years after Aristotle was shaped by his ideas. The concept of an Earth-centered universe was one of the most influential and long-lasting of all Aristotle's theories. It was part of the Christian Church's fundamental teaching until astronomers such as Nicolaus Copernicus and Galileo Galilei challenged it in the 15th and 16th centuries.

Aristotelian Logic

The term "logic" (from the Greek word *logos*, meaning speech or reasoning) was first used by Xenocrates (c. 395–314 BC), but it was Aristotle who provided some of the earliest surviving and complete texts on the subject. Aspects of Aristotelian logic continue to be taught today.

One of the most famous examples of Aristotle's logic is called the "Aristotelian syllogistic." A "syllogism" is an argument based on deductive reasoning. A syllogism must have two "premises," or assumptions, from which you can draw a conclusion. For example:

All birds are two-legged
All eagles are birds
All eagles are two-legged

The first statement is the "major" premise; the second is the "minor" premise; the last line draws a conclusion. The premises can be universal ("all," or "no") or particular ("some"); must be either affirmative (all As are B), or negative (no As are B); have a sign of quantity ("all"); and must contain three terms: major (in this case, "two-legged"), minor ("eagles"), and middle ("birds").

A wrongly constructed syllogism can result in a logically flawed argument or "fallacy," as, for example: All birds can fly; helicopters can fly; therefore helicopters are birds!

A B C

A universal, affirmative syllogism can be represented by three circles within one another. The major term encloses both the middle term and the minor term. Here we argue that all eagles (C) are birds (B), and so are two-legged (A).

▶ **see also** Copernicus VOL 1:14 Galileo VOL 1:24 Kepler VOL 1:38 Newton VOL 1:54

ARISTOTLE: Life and Times

SCIENTIFIC BACKGROUND

Before 390 BC

Greek philosophers make studies of the natural world; Democritus (c. 460–c.370 BC) develops the theory of atomism—that the world is composed of indivisible entities called atoms

Hippocrates of Cos (?c. 460–377 or 359 BC) lays the scientific foundations of medicine

390 BC

c. 387 Greek philosopher Plato (c. 428–c. 348 BC) founds his Academy in Athens; develops theory of forms; studies cosmology (the origin and nature of the universe) and physiology (the processes and functions of living animals and plants)

380 BC

370 BC

c. 370 Eudoxus of Cnidus (408–353 BC) proposes a new model of planetary motion; he accurately calculates the length of the solar year

360 BC

350 BC

345–42 Aristotle studies natural history on Lesbos

340 BC

336 Aristotle begins work on a comprehensive survey of all knowledge

335 Aristotle founds the Lyceum in Athens; he records his lectures in manuscript form, organizes research, and teaches

330 BC

322 Theophrastus (c. 372–c. 287 BC) succeeds Aristotle as head of the Lyceum and continues Aristotle's work on natural science, producing important works on plants

320 BC

310 BC

300 BC

After 300 BC

9th–10th centuries Aristotle and other Greek scientists translated into Arabic; **12th–13th centuries** Aristotle's surviving works translated into Latin

1543 Aristotle's Earth-centered view of the cosmos challenged by Nicolaus Copernicus (1473–1543) and Galileo Galilei (1564–1642)

POLITICAL AND CULTURAL BACKGROUND

399 The Greek philosopher Socrates (469–399 BC) is sentenced to death by drinking hemlock, a deadly poison

390 The Celts, or Gauls, a group of peoples from central and western Europe, invade and plunder Rome, the center of the Roman republic

c.385 Death of Aristophanes (born c. 448 BC), one of the most prolific of Greek playwrights, said to have written 54 plays, of which only 11 survive

359 Philip II (382–336 BC) becomes King of Macedonia

356 The temple of Artemis at Ephesus (in modern-day western Turkey) is burned down; it was one of the seven wonders of the ancient world

336 Philip II is assassinated; His son Alexander (356–323 BC) succeeds to the Macedonian throne

332 Alexander takes over Egypt and founds the city of Alexandria; he becomes known as Alexander the Great

331 Alexander the Great conquers the Persian empire and goes on to invade northern India

323 Alexander the Great dies, aged 32

323 The great museum at Alexandria is founded by Ptolemy (d. 283 BC), a former general in Alexander's army, who becomes governor of Egypt

312 The first aqueduct in Rome is completed to bring a supply of pure drinking water into the city

305 Ptolemy begins ruling in Egypt as Ptolemy I and founds a new dynasty of pharaohs

Nicolaus Copernicus

1473–1543

"At rest, however, in the middle of everything, is the Sun. For in this most beautiful temple, who would place this lamp in another or better position than that from which it can light up the whole thing at the same time? For the Sun is not inappropriately called by some people the lantern of the universe...."

Nicolaus Copernicus
On the Revolutions of the Celestial Spheres
(1543)

BY CHALLENGING THE ANCIENT IDEA THAT THE EARTH LAY AT THE CENTER OF THE UNIVERSE, COPERNICUS REVOLUTIONIZED ASTRONOMICAL BELIEFS AND HAS BECOME KNOWN AS THE FOUNDER OF MODERN ASTRONOMY. HIS THEORIES AROUSED HOSTILITY AMONG PROTESTANT CHURCH LEADERS.

Nicolaus Copernicus was born in 1473 in Torún, a town on the banks of the River Vistula in eastern Poland, where his father was a wealthy merchant. At the age of 18 Copernicus went to study at the University of Kraków, and then traveled to Italy, where he spent most of the following 10 years.

Italy was a very important center of learning in medieval Europe; students traveled there from all over Europe. The university of Bologna was Italy's oldest; it had been founded in the 11th century. It quickly became the most important center in Europe for the study of law and also provided teaching in medicine and philosophy. In 1222 Italy's second university was established in Padua; others, including Ferrara in 1391, followed. Copernicus studied philosophy, mathematics, and astronomy at all three universities during his years in Italy, and also qualified as a canon lawyer (one specializing in Church law) and physician.

On his return to Poland in 1506, Copernicus served as secretary to his uncle Lucas, who was Bishop of Ermland, and lived at the bishop's official residence of Heilsberg Castle. After his uncle died in 1512, Copernicus moved to Frauenburg in East Prussia, where he was appointed a canon of the cathedral. A canon is one of the priests responsible for organizing cathedral services and looking after the building. It was not a very demanding post, and it allowed him ample time to pursue his study of astronomy.

Ptolemy's System

The Greek philosopher Aristotle had described his view of the universe in the 4th century BC, in which transparent spheres carried heavenly bodies in perfectly circular orbits around an immobile Earth. However, astronomers watching the heavenly

A painting by the 19th-century Polish artist Jan Mateijko shows Copernicus, surrounded by astronomical instruments, in his roofless observatory. He built it on the fortified walls of Frauenburg Cathedral, which can be seen in the background.

bodies could see that the Sun, the Moon, and the planets did not actually move in this way. To allow for this, the astronomer Ptolemy (c. 90–168 AD) had to introduce a number of complications into the system. One of his proposals was that while the Earth remained at the center of a planet's orbit, the planet moved in another cycle, called an epicycle,

KEY DATES	
1473	Born on February 19 in Torún, Poland
1483	Father dies; adopted by his uncle Lucas Watzenrode
1496–1506	Studies law, medicine, and astronomy at the universities of Bologna, Padua, and Ferrara in Italy
c. 1510–14	Works on manuscript in which he theorizes that the Sun is at the center of the universe
1533	Presents his theory of a Sun-centered (heliocentric) universe in a lecture to Pope Clement VII
1543	Finally publishes heliocentric theories in his book *On the Revolutions of the Celestial Spheres*
1543	Dies on May 23 at Frauenburg

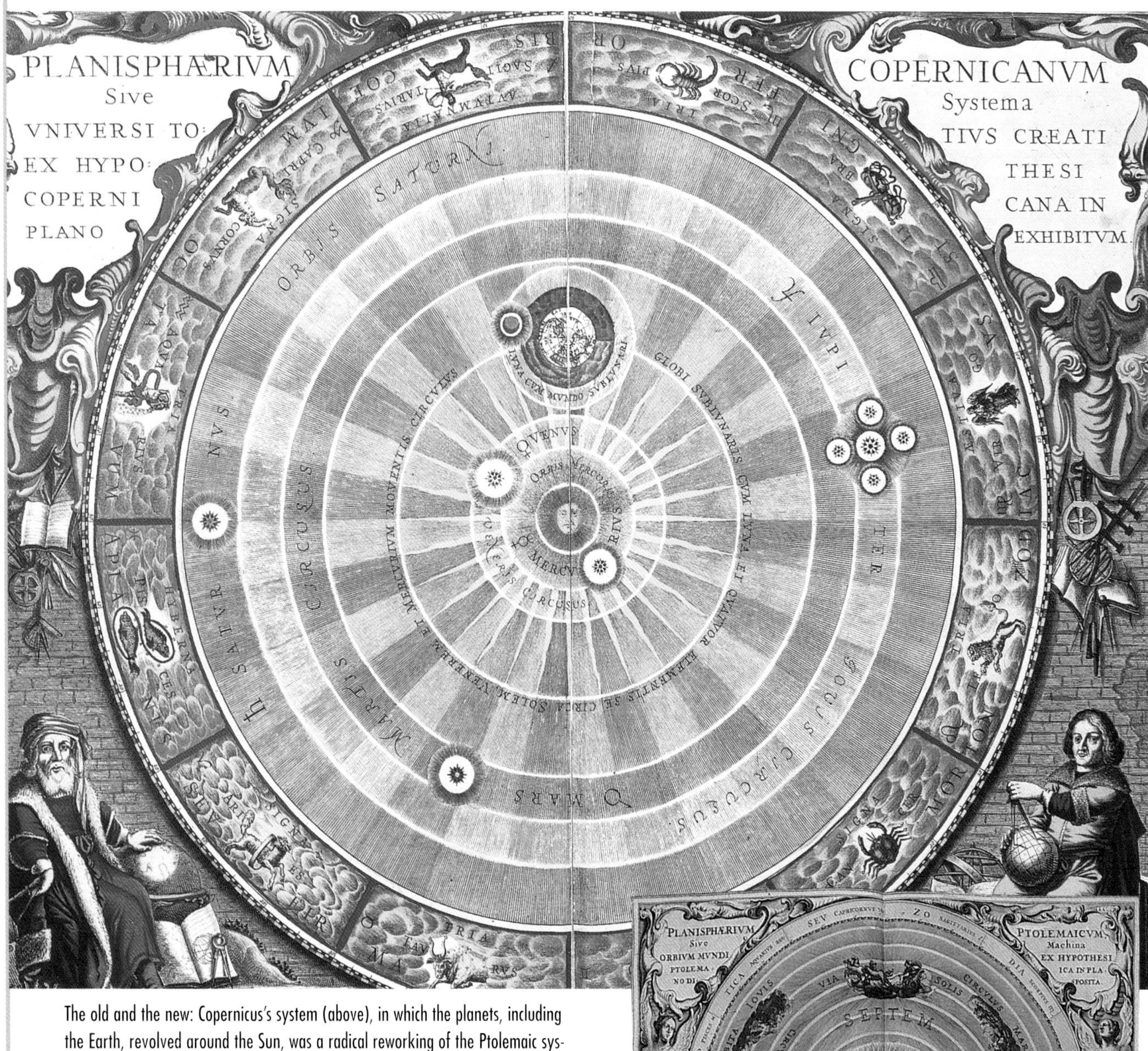

The old and the new: Copernicus's system (above), in which the planets, including the Earth, revolved around the Sun, was a radical reworking of the Ptolemaic system (right) in which the Earth was central and surrounded by water, air, and fire.

centered on its own orbit, or "deferent" (see diagram opposite). This was the only way Ptolemy could justify Aristotle's view that the heavenly bodies moved at uniform speeds throughout their orbits. More than 1,000 years later this system, called the Ptolemaic system, was still accepted as essentially correct.

"Sun, Stand Thou Still"

There appeared to be good reason for taking Ptolemy's view of the universe as true. Firstly, the Sun does seem to move across the sky each day, rising in the east, climbing and descending in a slow arc, and finally setting in the west. To an observer the Sun, rather than the Earth, appears to be moving. Secondly, astronomers reasoned that the Earth could not rotate on its own axis (a view held by some ancient Greek philosophers, including

Philolaos [c. 530 BC]) because if it did, why didn't buildings collapse or people feel a tremendous rush of wind against their faces? Why did an apple falling from a tree land at the tree's foot and not behind it? In the time it took the apple to fall to the ground, wouldn't the Earth have moved around some distance, taking the tree with it?

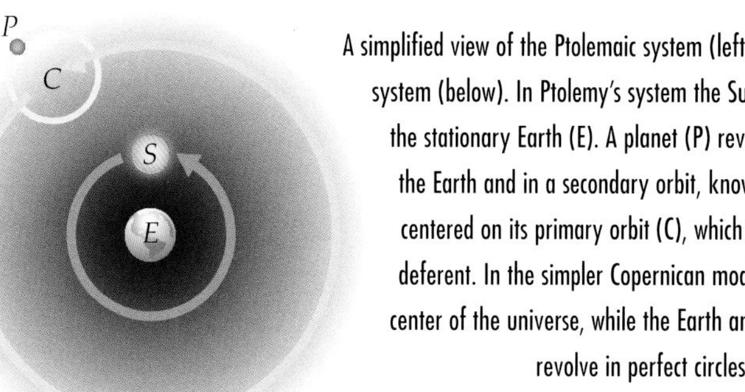

A simplified view of the Ptolemaic system (left) and the Copernican system (below). In Ptolemy's system the Sun (S) revolves around the stationary Earth (E). A planet (P) revolves both around the Earth and in a secondary orbit, known as an epicycle, centered on its primary orbit (C), which Ptolemy called its deferent. In the simpler Copernican model the Sun lies at the center of the universe, while the Earth and the other planet (P) revolve in perfect circles around it.

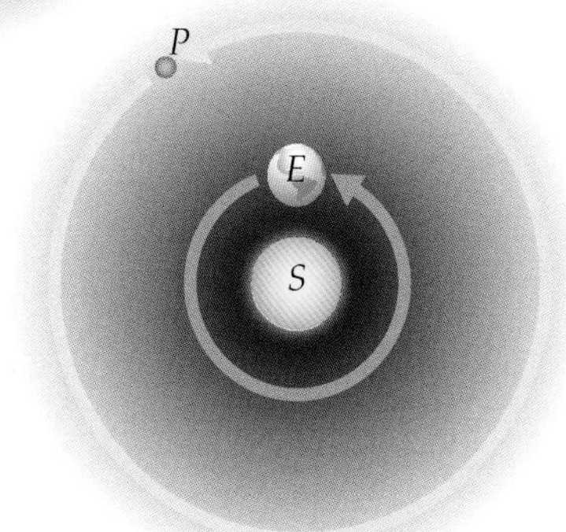

A third argument in favor of the idea that the Earth was stationary and the Sun moved was based on a famous passage in the Bible. The prophet Joshua had commanded: "Sun, stand thou still..."(Joshua 10: 12–14), and to Christian thinkers in the Middle Ages this suggested that the Sun moved at God's will.

The Copernican Revolution

Copernicus was the first astronomer since the time of the Greeks to challenge these views. He identified a number of problems with the Ptolemaic system, in particular Ptolemy's belief that the planets moved in one circle around the equant while maintaining a uniform speed in another circle around the Earth. Copernicus thought that this was altogether too complicated. He was in favor of a simpler system, in which bodies moved in a uniform circular motion around a single point. Consequently, in about 1510, he began to explore other possibilities that would fit the available evidence.

Copernicus now began to share the view of Aristarchus of Samos and some other Greek thinkers that the Earth circles the Sun rather than the other way round (see page 10). In other words, he thought that the universe is "heliostatic" (the word is from the Greek and means "stationary Sun"). In his book, *On the Revolutions of the Celestial Spheres* (which was not published until the year he died, 1543), Copernicus pointed out that, whether it is the Earth or the heavens that are moving, the results will seem the same to an observer. We could be situated on an Earth that is standing still while the stars revolve around us, or on an Earth that performs a complete orbit every 24 hours; in both cases, the stars would appear to revolve around us. We have all been on a moving ship or in a traveling car that we think is standing still. We believe that the other ships or vehicle we see are moving when in fact they are stationary. So, Copernicus argued, it is with the Earth and the stars and Sun.

Copernicus decided that all the planets, including the Earth, moved around the Sun, and that the Moon revolved around the Earth. Copernicus believed that this

Ptolemy (Claudius Ptolemaeus)
c. 90–168 AD

Ptolemy was an astronomer and geographer who worked in Alexandria, Egypt, in the 2nd century AD. Almost nothing is known about his life, but he was the author of several great works. In his *Guide to Geography* he tried to chart the known world; he also wrote on optics in *Optica*. The art of mapmaking was one of several other subjects that he tackled.

Ptolemy's most famous work was his *Mathematical Collection*, later called *The Greatest Collection* or *Al Majisti* in Arabic. The Arabic name was corrupted to *Almagest*, and this is the name by which the work is still known. The text summarizes work carried out by ancient astronomers, and also presents new work by Ptolemy; the astronomy described in the book became known as the Ptolemaic system. In this the Earth stood at the center of the universe, surrounded by the eight crystal spheres described by Aristotle. Each sphere carried a different heavenly body—the Sun, Moon, and the five known planets; the eighth carried the stars. It was this system that Copernicus questioned.

would explain much more simply why the other planets moved in the way they did, and why there was a variation in the brightness of the planets. Copernicus was even able to place the planets in order from the Sun: Mercury was nearest, then Venus, the Earth, Mars, Jupiter, and Saturn (the other planets were still undiscovered at this date). Copernicus also took up the idea first argued by some earlier Greek astronomers that the Earth spins, or rotates, daily on its axis.

Copernicus believed that this arrangement fitted in more accurately with how the planets appeared to move, and it was more simple than the Ptolemaic system. However, Copernicus continued to accept the Aristotelian idea that there were eight transparent spheres carrying the five known planets, the Sun, the Moon, and the stars, and that the planets moved in perfect circular motion.

COPERNICUS AND THE CHURCH

Copernicus completed *On the Revolutions of the Celestial Spheres* in 1514, but did not publish it until 1543. The reason for this delay is not clear. It is possible that he might have been concerned about the reaction of the Church. At this time in Europe the Catholic Church was still extremely powerful; in 1551 the pope, the leader of the Catholic Church in Rome, declared that people should not just follow reason but should confirm "their opinions with the Holy Scripture, Traditions of the Apostles, sacred and approved Councils, and by the Constitutions and Authorities of the holy Fathers." By this he meant that people should make sure that their opinions were in line with those of the Catholic Church. So there did appear to be some risks for Copernicus if he published his theory because it contradicted a biblical text and suggested that humankind, which was supposed to have been created in God's image, did not stand at the very center of the universe. Many years later, in 1633, the Italian astronomer Galileo Galilei (1564–1642) would experience the full force of the Church's anger when he was put on trial for supporting Copernicus's theory (see box pages 30–31).

However, nearly a century before Galileo's trial it was the Protestant reformer Martin Luther (1483–1546) who seemed most horrified by Copernicus's new theories. Luther was leader of

Measuring the World

In the days before complex measuring instruments and satellites, astronomers had to rely on simpler methods. In the 3rd century BC Greek astronomer Eratosthenes (c. 276–194 BC) worked out how big the Earth was using just a pole, or "gnomon." He knew that at Syene (now Aswan), in Egypt, no shadow was cast by a gnomon at midday on midsummer's day because the Sun was then directly overhead. When he measured the shadow cast by a gnomon at Alexandria at exactly the same time he found that the Sun's rays fell at an angle of 7° from the vertical. Once he had calculated the distance between Alexandria and Syene (about 500 miles [800 km]), he was able to compute the circumference of the Earth to within about 130 miles (209 km) of the correct figure.

By Copernicus's day astronomers had a range of devices to help in their observations. The armillary sphere, a globe of the heavens, consisted of a number of calibrated metal rings representing the celestial equator, horizon, and so on. These allowed astronomers to calculate the position of the stars. You can see one in front of Galileo in the top picture on page 35. The plane astrolabe had a disk with a movable chart of the heavens and a pointer for measuring angles. The quadrant, consisting of a quarter circle marked in degrees with a movable arm to measure the altitude of stars, was convenient to use. Astronomers continued to rely on them to work out the position of stars and planets even after celestial telescopes were improved in the early 17th century.

Astrolabes were used by medieval scientists to work out the position of heavenly bodies, and by sailors to help them plot their position. They had a circular disk marked in degrees and a movable pointer.

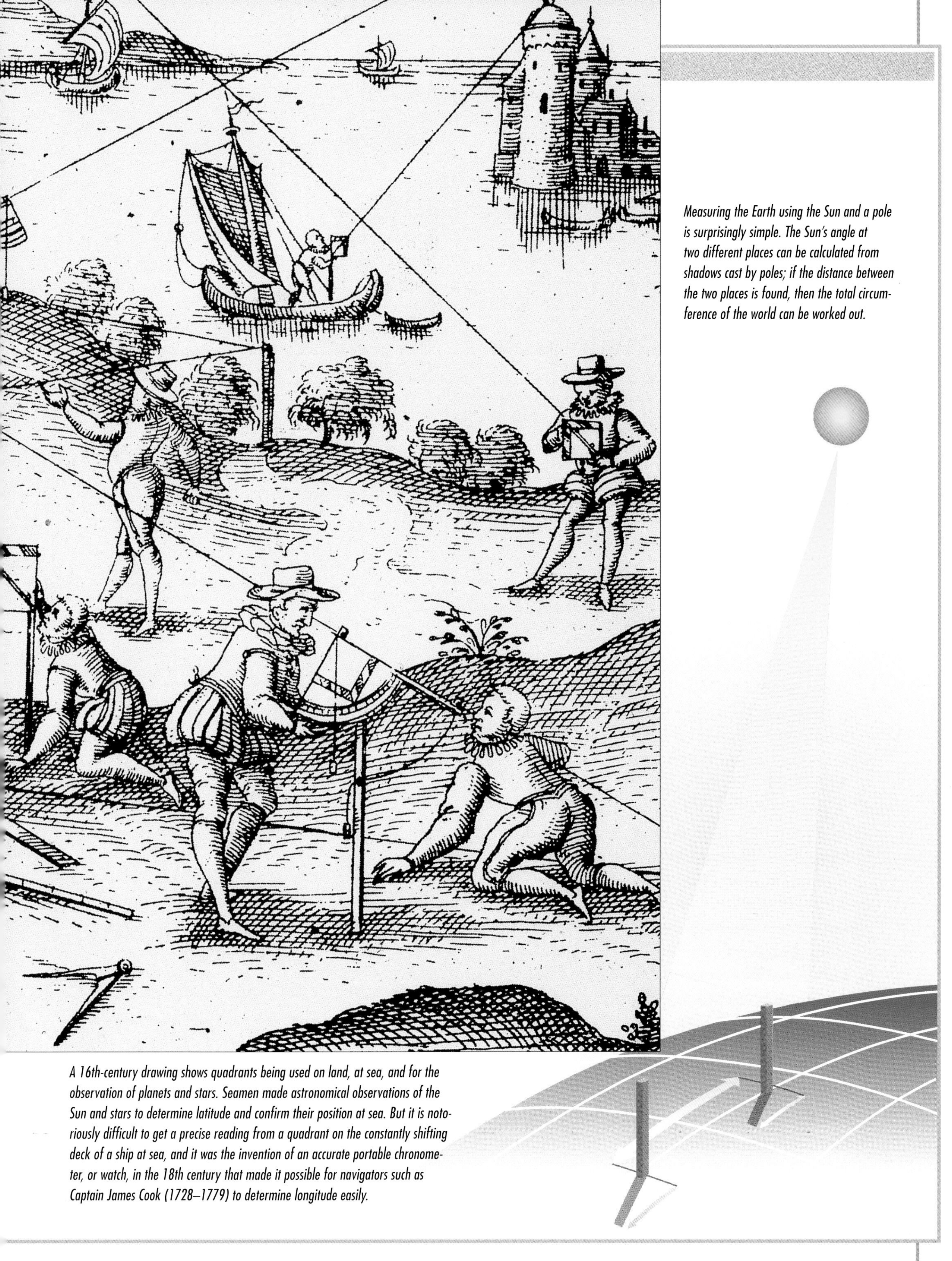

Measuring the Earth using the Sun and a pole is surprisingly simple. The Sun's angle at two different places can be calculated from shadows cast by poles; if the distance between the two places is found, then the total circumference of the world can be worked out.

A 16th-century drawing shows quadrants being used on land, at sea, and for the observation of planets and stars. Seamen made astronomical observations of the Sun and stars to determine latitude and confirm their position at sea. But it is notoriously difficult to get a precise reading from a quadrant on the constantly shifting deck of a ship at sea, and it was the invention of an accurate portable chronometer, or watch, in the 18th century that made it possible for navigators such as Captain James Cook (1728–1779) to determine longitude easily.

the attack on corruption in the Roman Catholic Church that resulted in the movement known as the Reformation. The Reformation led to a break with the pope and the establishment of the Protestant faith. Pointing out that "sacred Scripture tells us that Joshua commanded the Sun to stand still, and not the Earth," Luther dismissed Copernicus in no uncertain terms as an upstart astrologer and a fool. His fellow reformer Philip Melanchthon (1497–1560) added to Luther's objections by commenting that, "...It is want of honesty and decency to assert such notions publicly, and the example is pernicious."

> ***"It is want of honesty and decency to assert such notions publicly, and the example is pernicious..."***
>
> PHILIP MELANCHTHON, LUTHERAN REFORMER, ON COPERNICUS'S THEORIES

FEAR OF RIDICULE

Copernicus had already presented his theory of a Sun-centered universe in a lecture to Pope Clement VII in 1533, so in fact his reluctance to publish his ideas seems to have had more to do with his concerns about people misunderstanding his arguments than with threats from the Church. Copernicus claimed that he had deliberately made his work as technical as possible so that it could be judged only by mathematicians, insisting that, "Mathematics is for mathematicians." He also added to the title page of his book the motto taken from Plato's Academy (see box page 7): "Let no one enter who knows no Geometry."

A PERSUASIVE PROFESSOR

Copernicus might never have published *On the Revolutions of the Celestial Spheres* had it not been for the young Georg Joachim von Lauchen (1514–1574), an Austrian-born scholar who is better known by the Latin name of Rheticus. Rheticus had learned of Copernicus's theory of a Sun-centered universe in Wittenberg, Germany, where he was professor of mathematics. Coincidentally, Wittenberg was also where Martin Luther was a professor. The Protestant Reformation had begun in Wittenberg when Luther pinned to the church door there articles—known as the Ninety-Five Theses—that were highly critical of the pope and the clergy.

In 1539, aged only 25, Rheticus arrived unannounced at Copernicus's home. He stayed there for nearly two years. It was a surprising alliance that developed between the two: Rheticus was an energetic and ambitious young man, and Copernicus was an aging and retiring priest. In addition, Rheticus was a Protestant visiting a Catholic at a time in the early Reformation when religious differences could cost a man his job, his freedom, and even his life.

During this time Rheticus set about mastering the Copernican system of the universe, and in 1540 he published a brief description of it in *First Account of the*

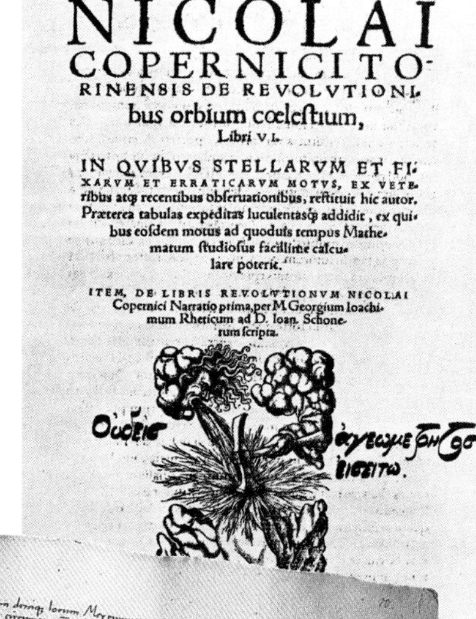

NICOLAI COPERNICI TORINENSIS DE REVOLVTIONIbus orbium cœlestium, Libri VI.

IN QVIBVS STELLARVM ET FIXARVM ET ERRATICARVM MOTVS, EX VETEribus atq; recentibus obseruationibus, restituit hic autor. Præterea tabulas expeditas luculentasq; addidit, ex quibus eosdem motus ad quoduis tempus Mathematum studiosus facillime calculare poterit.

ITEM, DE LIBRIS REVOLVTIONVM NICOLAI Copernici Narratio prima, per M. Georgium Ioachimum Rheticum ad D. Ioan. Schonerum scripta.

Οὐδεὶς ἀγεωμέτρητος εἰσίτω.

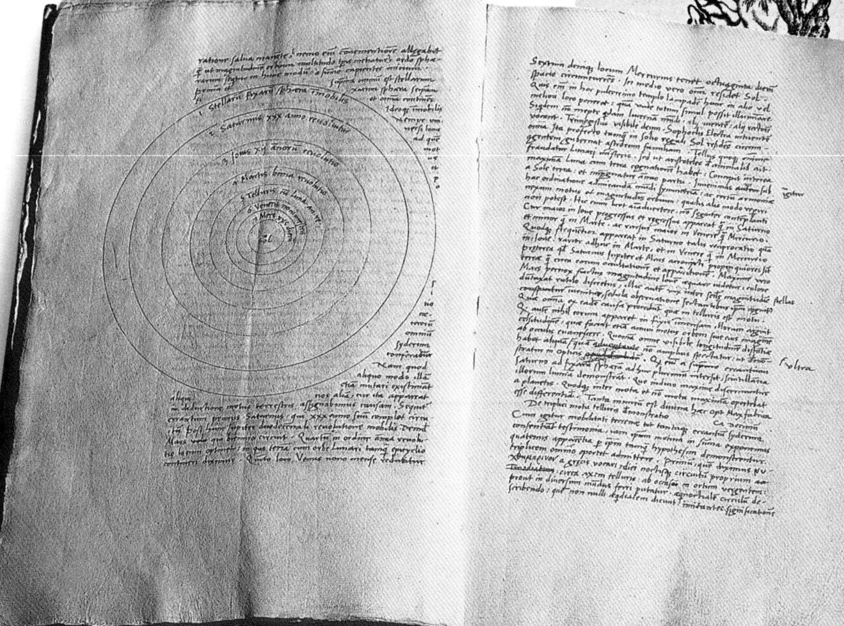

The title page (right) and manuscript (below) of Book I of *On the Revolutions of the Celestial Spheres*. In this book Copernicus argued against an Earth-centered universe, and placed the known planets in order around the central, stationary Sun.

Reforming the Calendar

In 1514 Pope Leo X (1475–1521) invited Copernicus to Rome to advise the Church on revising the calendar. The existing calendar had been established in the 1st century BC by the Roman statesman Julius Caesar (100/102–44 BC), and was named the Julian calendar for him. The Julian calendar had replaced the old lunar calendar (one based on the phases of the Moon). In reforming the calendar, Caesar called on the help of the Alexandrian astronomer Sosigenes, and it was Sosigenes who suggested basing the new calendar on a solar calendar (one related to the Sun), with a year length of 365 1/4 days. He then divided this into months based on the seasons. To achieve an average year length of 365 1/4 days, it was decided that three years should be 365 days long, and the fourth should be 366 days: this extra day still appears in the modern calendar every four years, or leap year, as February 29. There was a 90-day difference between the old lunar calendar and the new Julian one, which came into use in 46 BC. To make up this gap, officials inserted 23 extra days after February 23 and 67 extra days at the end of November. That year was 445 days long.

The new calendar seemed to be much more accurate; the fact that each year was just over 11 minutes too long did not seem to matter. However, as the years went by the seasons again became increasingly out of phase with the calendar, leading to the pope's invitation to Copernicus to revise it. He turned it down, arguing that it was important to work out the proper motions of the solar system before attempting any revisions. By 1545 the spring equinox, used to calculate the date of Easter, was 10 days adrift from the calendar date.

It was not until 1582 that attempts were again made to revise the calendar: Pope Gregory XIII (1502–1585) adjusted it by jumping straight from October 5 to 15. Now the length of the year was accepted as 365.2422 days, which differed from the Julian calendar by 3.12 days every 400 years. Leap years were still retained every four years. But to make sure the calendar stayed in phase with the solar year it was decided that centennial years (those divisible by 100 as well as by 4) should not be kept as leap years. The only exceptions were quatercentennial years (those divisible by 400). That is why the year 2000 was a leap year but 1700, 1800, and 1900 were not.

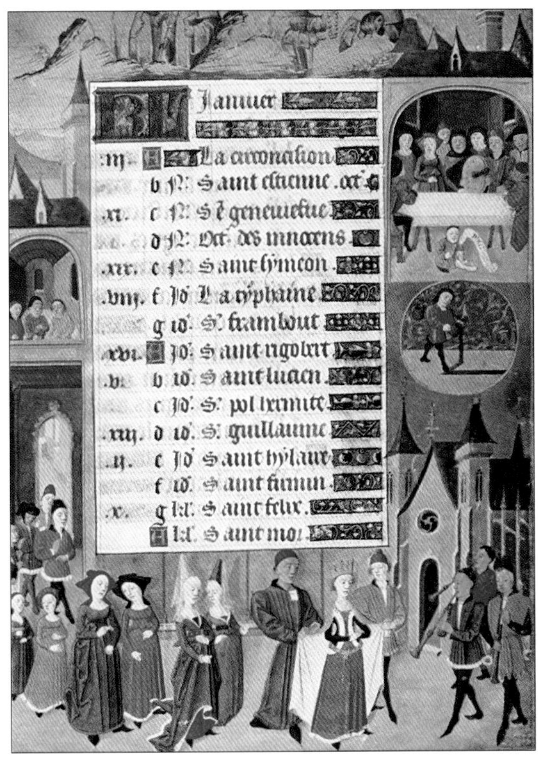

A 15th-century illustration of the month of January shows a coronation in southern France. The Julian calendar was the first to introduce months based on seasons; from 1582 it was replaced by the Gregorian calendar.

The new calendar, known as the Gregorian, was adopted throughout Catholic Europe in 1582. Britain and America did not begin using it until 1752. Other countries were slower still; the Soviet Union adopted it only in 1918 and Greece in 1923.

Book On the Revolutions by Nicolaus Copernicus. Eventually Rheticus managed to persuade Copernicus to have his manuscript printed, and to allow it to be circulated to a wider public.

THE NEW ART OF PRINTING

Printing had been invented in Mainz, Germany, about a century earlier. Known popularly as the German art, it spread rapidly to all the major centers of commerce and learning. By about 1470 Nuremberg, in southern Germany, had became the chief center of a flourishing book publishing industry. It was here that Copernicus's text was taken for publication.

Unknown to Copernicus, an addition was made to his text by the priest Andreas Osiander (1498–1522). Osiander was a supporter of Luther, and had been shown the text by Rheticus. Osiander inserted a preface saying that Copernicus's theory was only a way of tying in calculations with observations, and that it should not be taken literally. This disclaimer might have made the work more acceptable to the public. By the time it was ready for publication Copernicus had suffered a stroke. A first copy of the printed book is said to have been brought to him at Frauenberg on the day he died.

SHIFTING STARS

Copernicus's ideas were now revealed to other astronomers, the Church, and the public. The most important astronomical criticism came from the Danish astronomer Tycho Brahe (1546–1601). Brahe (see box page 40) argued that if the Earth were orbiting the Sun once a year as Copernicus claimed, then anyone observing the stars on a regular basis would view them from widely different observational points during the course of the year, and so should expect to see a shift in the pattern of the stars. Because astronomers of the time could not see any such shift, they decided that the Earth could not be moving. In fact, the shift does take place. It is called annual parallax. It is similar to the effect that can be achieved if you close one eye and focus on a distant object; if you then open that eye and close the other one, you will notice that the object shifts slightly.

Copernicus had already considered this point. He was convinced that the parallax could not be seen because the stars were so far away. This was another point of dispute; many astronomers of the time could not imagine that God would design a universe in which such a huge space stood between the stars and the planets. Copernicus was correct, though. There is annual parallax, but it was not detected until 1838 when the German astronomer Friedrich Bessel (1784–1846), was able to detect a slight degree of parallax in the star 61 Cygni, about 10.3 light-years from Earth (the nearest known star, Proxima Centauri, is about 4.26 light-years away).

Georg Joachim von Lauchen (known as Rheticus) 1514–1574

Rheticus was born in 1514 in Feldkirch, Austria. In 1536 he was appointed professor of mathematics and astronomy at Wittenberg. There he heard of Copernicus's theory of a Sun-centered universe and, fascinated by the idea, traveled to Poland to visit and study with Copernicus. After persuading Copernicus to publish his theory, Rheticus later produced his own great work, *The Palatine Work on Triangles*, which was completed after his death by his pupil Valentin Otto and finally published in 1596. The book contains trigonometric tables to calculate arcs and angles.

The same group of stars viewed from Earth at two different times of the year will seem to alter slightly in pattern (boxes A and B). This apparent shift, due to the Earth's orbiting of the Sun, is called annual parallax. The two different positions of the observer form two points of the triangle, and the position of the stars the third. Copernicus correctly asserted that annual parallax could not be detected with the astronomical instruments of his day because the stars were so far away.

◀ **see also** Aristotle VOL 1:6
▶ **see also** Galileo VOL 1:24 Kepler VOL 1:38

COPERNICUS: Life and Times

SCIENTIFIC BACKGROUND

Before 1480

Greek philosopher Aristarchus of Samos (c. 310–230 BC) teaches that the Earth orbits a stationary Sun, but his theory is not widely accepted

Astronomer Ptolemy of Alexandria (c. 90–168 AD) sums up Greek astronomy, principally that the Earth is at the center of the cosmos

1480

1483 The *Alfonsine Tables*, a revision of the *Ptolemaic Tables* of planetary positions, are printed in Toledo, Spain

1490

1497 Copernicus observes and records the temporary disappearance of a star behind the Moon

1500

1504 Columbus frightens a group of Native Americans by correctly predicting a total eclipse of the Moon on February 29

1510

1514 Copernicus writes the first version of his heliocentric (Sun-centered) theory, but does not publish it for nearly 30 years

1520

1530

1540

1540 The Austrian-born mathematician and astronomer Rheticus (1514–74) publishes his *First Account* of Copernicus's heliocentric theory

1543 Copernicus publishes his heliocentric theory in *On the Revolutions of the Celestial Spheres*

1550

After 1550

1577 Danish astronomer Tycho Brahe (1546–1601) shows that comets move in spaces between planets, and do not lie within the Earth's atmosphere

1609 German astronomer Johannes Kepler (1571–1630) shows that planetary motion is elliptical, and that the speed of a planet's orbit speeds up when it is nearer the Sun and slows down when it is farther away

1616 Copernicus's book is placed on the Catholic *Index of Prohibited Books*

POLITICAL AND CULTURAL BACKGROUND

1474 English printer William Caxton (c. 1422–c. 1491) prints the first ever book in the English language, *Recuyell of the Historyes of Troye*

c. 1484 The High Renaissance blooms in Italy; Italian artist Sandro Botticelli (1444–1510) paints his mythological work, *The Birth of Venus*

1492 In a period of great exploration, Genoese sailor Christopher Columbus (1451–1506) becomes the first European to discover the New World

1497 Genoese-born Venetian explorer John Cabot (Giovanni Caboto, 1425–c. 1500) sights North America, claiming the land for England

c. 1504 Italian painter Leonardo da Vinci (1452–1519) completes his most famous picture, the *Mona Lisa*

1508–12 Italian painter and sculptor Michelangelo (1475–1564) paints the Sistine Chapel ceiling in Rome

1517 German religious reformer Martin Luther (1483–1546) instigates the Protestant Reformation by displaying articles critical of the Catholic Church on the castle church door at Wittenberg

1519 Spanish conquistador Hernando Cortés (1485–1547) begins his conquest of the Aztec empire in Mexico on behalf of Spain

1526 The Mogul dynasty of Muslim emperors is established in India; Moguls will rule there for more than 330 years

1533 Aged three, Ivan IV (1530–84) assumes power of Russia; his savage rule earns him the name "Ivan the Terrible"

1534 English King Henry VIII (1491–1547), who wants to divorce his first wife and remarry, breaks with Roman Catholic Rome and heads a newly established Church of England

1545–63 The Council of Trent attempts to define doctrine and reform abuses in the Roman Catholic Church; it marks the start of a Counter-Reformation against Protestants

GALILEO GALILEI

1564–1642

"He hath first overturned all former astronomy... and next all astrology."

The English ambassador to Venice, on hearing of the theories Galileo published in The Starry Messenger (1610)

GALILEO GALILEI WAS A MATHEMATICIAN AND ASTRONOMER WHOSE METHODS MADE HIM THE FIRST TRULY "MODERN" SCIENTIST. HE MADE HIGHLY SIGNIFICANT CONTRIBUTIONS IN THE FIELDS OF GRAVITATION, MOTION, AND ASTRONOMY, BUT HE CAME INTO CONFLICT WITH THE CATHOLIC CHURCH OVER HIS VIEWS ABOUT THE UNIVERSE.

GALILEO GALILEI WAS BORN IN 1564 IN PISA, IN the region of Tuscany in Italy, which was then ruled by the Grand Duke Cosimo de' Medici, known as Cosimo I. Galileo was the eldest of seven children. His father, Vincenzio Galilei, was a distinguished musician—a lute player, who wrote some important works on musical theory. Galileo attended a school run by the Jesuit religious order in Florence, the capital city of Tuscany, and in 1581 enrolled at the University of Pisa to study medicine. This could have led to a well-paid position, but the subject failed to stimulate him. Galileo's real interest lay in mathematics, and he began to devote all his spare time to it, becoming a lecturer in mathematics at the university in 1589. He gained a reputation as an original thinker, and was considered disrespectful and disruptive by many.

Throughout his life Galileo was constantly short of money. With the death of his father in 1591 he was obliged to provide dowries for his sisters, Virginia and Livia, who married in 1591 and 1601 respectively. Dowries were payments of money or property that were made by a woman's family to her husband. They could be paid in instalments over many years, but failure to meet a payment could lead to arrest and imprisonment. To avoid this Galileo borrowed money, and spent much of his time teaching private students.

In 1592 Galileo moved to Padua; it was a larger university offering better facilities, a more tolerant atmosphere, and increased pay. While there he met and formed a relationship with Marina Gamba, and they had three children. However, when Galileo returned to Florence 18 years later, in 1610, Marina stayed behind in Padua and eventually married someone else. Because they were illegitimate, their two daughters had no prospect of marriage and so they became nuns.

KEY DATES

1564	February 15, Galileo born in Pisa, Italy
1574	Moves to Florence with his family and attends a Jesuit school there
1581	Enters the University of Pisa to study medicine
1589	Appointed lecturer in mathematics at the University of Pisa
1590	Criticizes Aristotle's ideas on motion in his treatise, *Motion*
1592	Appointed professor of mathematics at the University of Padua
1593	Invents early form of thermometer
1609	Constructs improved telescope; uses it to observe the heavens
1610	Returns to Florence as mathematician to the grand duke of Tuscany; publishes *The Starry Messenger*
1632	*Dialog on the Two Chief Systems of the World* published
1633	Called before the Inquisition in Rome, interrogated, threatened with torture, and forced to withdraw his theories of a Sun-centered universe; placed under house arrest at Arcetri
1638	*Discourses upon Two New Sciences* published in the Netherlands
1642	January 8, dies at Arcetri

Galileo was a practical scientist, who enjoyed performing experiments. After hearing about a new instrument, the telescope, he set about constructing his own. From 1609 it became an invaluable tool in his efforts to understand the cosmos. This 19th-century engraving of Galileo shows him at work in his study.

AN EXPERIMENTAL APPROACH

Galileo took a much more practical approach to his work than scientists before him had done, opting to perform experiments rather than to deduce facts through theory alone. While many scientists published the results of their work in scholarly texts written in Latin, Galileo chose to present his results in everyday Italian, and in lively, readable prose that was guaranteed to reach a wider public.

While he was at Padua Galileo began his investigations into motion. He was able to prove that a projectile such as an arrow shot from a bow, or a cannon ball from a cannon, does not travel in a straight line and then drop to the ground as the Greek philosopher Aristotle (384–322 BC) had believed (see page 11), but that it travels in a curve known as a parabola. Galileo had been wrestling with other problems of motion, too, and he now needed to test his theory about falling objects. Aristotle's view had been that the rate at which an object falls is related to its weight, so that a large, heavy rock would fall faster than a small, light pea. If both were released from the same height, Aristotle argued that the heavier body would reach the ground first.

"About ten months ago a report reached my ears that a certain Fleming had constructed a certain spyglass by means of which visible objects though very distant from the eye of the observer were distinctly seen as if nearby."

GALILEO IN 1609, ON HEARING OF A NEW INVENTION, THE TELESCOPE

DROPPED FROM A HEIGHT

Galileo disagreed. He thought that all objects fall through the air at the same speed. According to tradition, he is supposed to have traveled to Pisa, a short distance from Padua, in 1591. There he carried two cannon balls, one weighing 10 pounds (4.5 kg) and the other 1 pound (0.45 kg), to the top of the 179-foot (54.3 m) Leaning Tower. He placed them on the overhanging edge and then, calling on the large crowd gathered below to take note of what happened, released them at precisely the same time. Both struck the ground simultaneously.

Although this story has been told of Galileo over the centuries, documentary evidence shows that the experiment was in fact made by a rival who hoped to prove Galileo's theories wrong. If he had dropped a cannon ball and a sheet of paper, air resistance would have kept the sheet of paper fluttering through the air for longer. However, the

The bell tower—better known as the Leaning Tower—at Pisa, Italy, as seen from the archbishop's palace. On the desk are Galileo's later manuscript on motion, *Discourses upon Two New Sciences* (1638), and two cannon balls similar to those he is supposed to have dropped from the tower in his 1591 experiment.

Robert Boyle
1627–1691

Robert Boyle was born in County Waterford, Ireland, the 14th child of a wealthy English family. In 1660, while living in Oxford, England, he carried out a number of experiments on air resistance. His investigations—using a bullet (which passes easily through air) and a feather (which has a large surface area so encounters more air resistance)—showed that in a vacuum two objects fall at the same speed, despite being of different structures and weights. These results confirmed what Galileo had found in his earlier experiments on free-falling objects.

two cannon balls, despite their different weights, were of similar shape, and so plummeted to the ground with equal force.

A New Invention

Galileo soon came to accept the theory put forward by Nicolaus Copernicus (1473–1543) that the planets revolved around the Sun (see page 17), but he was reluctant to make his views public. Things changed once he heard news of a recent invention that had been made in Flanders (modern Belgium): "About ten months ago," Galileo wrote in around 1609, "a report reached my ears that a certain Fleming [an inhabitant of Flanders] had constructed a certain spyglass [a telescope] by means of which visible objects though very distant from the eye of the observer were distinctly seen as if nearby."

Galileo was intrigued by the report and set about trying to produce a similar instrument. He prepared a hollow lead tube and fitted two glass lenses at either end. One was concave (rounded inward on one side) and the other was convex (rounded outward on one side). He placed his eye to the concave lens and found that distant objects appeared three to nine times larger than when observed with the naked eye. He went on to construct an even more accurate instrument that enlarged distant objects more than 60 times their actual size.

Magnifying the Heavens

In late 1609 Galileo began to observe the heavens with his new telescope. While watching the planet Jupiter on January 7, 1610, he noticed something strange, and began to make sketches of his observations. On the first night he saw three objects close to Jupiter, two on one side and one on the other. The objects, which he took to be stars, lay in a straight line. The next night Galileo was surprised to see three objects to the west of Jupiter and none at all to the east. Two nights later two objects were visible to the east and none could be seen to the west.

After continuing these nighttime observations throughout the whole of January, Galileo came to the inescapable conclusion that there were three stars "wandering around Jupiter, like Venus and Mercury around the Sun." (He later revised this number to four.) As they circled the planet, they were sometimes lost from sight behind it, or sometimes visible from Earth to the east of Jupiter and sometimes to its west. Although Galileo referred to them as stars, we now know them as Jupiter's moons, or satellites.

Galileo was eager to demonstrate his discovery to other astronomers and philosophers. He took his new telescope to Bologna, where he invited his colleagues to look through it, but the demonstration was not instantly successful. Martin Horky, an astronomer from Bohemia (the modern Czech Republic), and a pupil of German astronomer

A page from Galileo's notebook (left) records the position of the "stars" of Jupiter over several days in 1612 and 1613. Two of the telescopes made by Galileo (below), through which he was able to observe the planets and discover the host of stars in the Milky War. Sky-watching with this new instrument became a popular pastime.

A 16th-century depiction of Padua University (left). It was founded in 1222 by students who had left the nearby Bologna University; the students selected the professors and paid their salaries. By the time Galileo arrived there in 1592 it was one of the leading two or three universities in Europe.

Johannes Kepler (1571–1630), claimed that the telescope worked well while observing earthly objects, but produced "fictions" when pointed at the heavens. Others proved unwilling even to look through the telescope. "Oh, my dear Kepler," Galileo wrote, "how I wish that we could have one hearty laugh together! Here at Padua is the principal professor of philosophy, whom I have repeatedly and urgently requested to look at the Moon and planets through my glass which he refuses to do."

Mountains on the Moon

Until this date, astronomers had believed that the heavenly bodies, including the Moon, were perfectly smooth and spherical. Observing the Moon through his telescope, Galileo made a startling discovery. He was the first person to recognize that there are mountains on the Moon. As he wrote in *The Starry Messenger*, published in 1610, he discovered that the Moon's surface was "uneven, rough, and crowded with depressions and bulges. And it is like the face of the Earth itself, which is marked here and there with chains of mountains and depths of valleys."

Galileo was amazed by the vast numbers of stars that his telescope revealed for the first time, "so many as to be almost beyond belief." He turned his attention to the galaxy, or Milky Way, which is visible as a hazy band of light in the night sky. Since ancient times, philosophers and scientists had puzzled over the precise nature of this spectacle. Galileo was able to settle their disputes once and for all, concluding that: "the galaxy is nothing else but a mass of innumerable stars planted together in clusters. Upon whatever part of it you turn the telescope, a vast crowd of stars immediately presents itself to view."

Paintings of the phases of the Moon by Galileo clearly show the "uneven, rough" surface that he was able to see using his telescope.

Anxious to ensure that his discovery of the moons of Jupiter would be properly rewarded, Galileo decided to name them "The Medicean Stars," in honor of the Medici rulers of Tuscany. In this way he hoped to make his reputation and earn a higher salary. He therefore dedicated the account of his discovery to the current ruler of Tuscany, Grand Duke Cosimo II (1590–1620). The flattery paid off: in 1610 Galileo was appointed principal mathematician to the duke and was able to arrange his return to Florence, the ducal capital.

Disputes and Debates

Increasingly, Galileo was seen to be challenging the centuries-old authority of Aristotle and the Greek philosophers, and his widely published views offended many. In 1612 Galileo became involved in an ill-tempered dispute about the properties, or characteristics, of ice. Galileo argued that ice floats on water because its density (volume and weight) is less than water, but his opponents, following Aristotle, said it was due to its wide, flat shape. A public debate was proposed but the argument became so unpleasant that Cosimo II ruled that his officials should not be seen to dispute in public; he insisted on the debate being continued in writing.

In 1613 Galileo became involved in another argument, this time with the German astronomer Christoph Scheiner (1579–1650), who was a Jesuit priest. Galileo was convinced that he had been the first to observe spots on the surface of the Sun in

The German astronomer Christoph Scheiner observes sunspots on the Sun by projecting the Sun's image onto a screen through a telescope. Scheiner thought that the spots were satellites orbiting the Sun, whereas Galileo was correct in thinking the sunspots were on the Sun's surface.

1611; now, two years later, he thought Scheiner was seeking to claim credit for the discovery. Equally upsetting to Galileo was the fact that Scheiner believed that the spots could not exist on the Sun itself as this would imply that the Sun was imperfect and changeable, whereas Aristotle had said that it was perfect and unchanging. Scheiner therefore decided that sunspots must be orbiting satellites, located somewhere in space between the Earth and the Sun.

To prove him wrong, Galileo pointed out that when the sunspots were at the outer edge of the Sun, they seemed to move more slowly than when they were at its center. He correctly reasoned that the change in their speed would make sense if the sunspots formed part of the Sun, but would be most unlikely if they existed independently.

In Trouble for His Views

Because of the strength of his views, Galileo was gathering a number of very dangerous enemies in Rome. Much of his work and writings challenged traditional Church doctrines, based as they were on Aristotelian theory, and this could not be tolerated. In 1633 he stood trial before the Inquisition and received a life sentence (see box pages 30–31).

Galileo's sentence, commuted to house arrest, was served in a villa at Arcetri, a little way outside Florence. He now had leisure to complete his scientific masterpiece, Discourses upon Two New Sciences (1638), in which he brought together the threads of previously unpublished experiments that had been interrupted by his telescopic studies.

He now returned to his investigations into falling objects. He had established that when dropped from the same height, objects fall equal distances in equal times, whatever their weight. But if the weight of an object does not affect the rate (or velocity) at which it falls, what does?

Galileo realized he needed to find some means of measuring the rate at which objects fall. This was not easy, given the speed at which they fall and the lack of accurate timekeeping devices. Sand timers,

Galileo on Trial

The leaders of the Protestant Reformation had thrown down a challenge to the Roman Catholic Church, criticizing its teachings and the worldly attitudes of the clergy. In response the Church set up the Council of Trent, which met in 1545 and issued strict guidelines on how the Church should interpret Scripture (biblical writings). It also set out to reform the behavior of the clergy and placed more power in the hands of the popes. These events launched the "Counter-Reformation," by which the Church aimed to win back the support lost to Protestantism. As part of its efforts, the Roman Inquisition was established in 1542. The Inquisition challenged anyone who threatened the Roman Catholic view of the world by putting forward Protestant or heretical views (ones at odds with the teachings of the Catholic Church). Galileo's views made him an object of suspicion.

Science and Scripture

Galileo was a committed supporter of Copernicus, and of the Copernican theory that the Earth moved around the Sun—a view that was contrary to the Church's teachings. During the period 1615–33 Galileo struggled to resolve contradictions raised by the conflict between Copernicanism and traditional Church teachings. A long-running and difficult discussion with the Church authorities followed.

The Church's teachings centered on the perfect nature of God and his creations, as set down in the Bible. Galileo thought he had a simple and sensible defense. Scripture, he insisted, could never lie. But, equally, neither could Nature, which owed as much to God as Scripture did. Galileo also argued that much of Scripture had been written to be easily understood; it often dealt with difficult and puzzling aspects of Nature by using metaphors (figures of speech) and simplifications. Galileo suggested that where there was conflict between Scripture and Nature, we should look again at Scripture to see if we had interpreted it correctly.

Cardinal Bellarmine (1542–1621), a Jesuit theologian, was the chief defender of the Church's position at this time. He was also a friend and admirer of Galileo, but in 1616 he had to tell Galileo that Pope Paul V (1552–1621) had banned the teaching of the heliocentric, or Sun-centered, system of the universe. Galileo realized that any public defense of Copernican theory would be seen as an open challenge to the authority of the Church, so for many years he chose to keep his views to himself and moved on to the consideration of other scientific conundrums. Things appeared to change, however, with the election in 1623 of another old friend, Maffeo Barberini (1568–1644), as Pope Urban VIII. In the same year Galileo published *The Assayer*, in which he discussed scientific method and experimentation. *The Assayer* was well received by the pope, who had it read to him at meals; when Galileo visited Rome in 1624, he was showered with honors and granted private audiences.

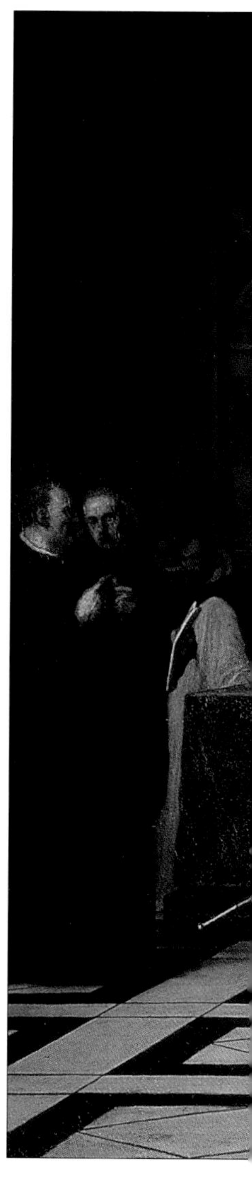

A Storm Breaks

All this led Galileo to believe that the climate of opinion was changing. At last, he felt, here was a pope who appeared sympathetic to his views. Even if he had enemies in Rome who were plotting against him, surely he could count on his friend, the pope, to shield him. In any event he was now living in the city-state of Florence, where he enjoyed the protection of the powerful and independent grand duke of Tuscany, Ferdinand II (reigned 1627–70). Galileo felt encouraged to seek permission to write an account of the universe. It was granted to him, but he was told he should come to a conclusion dictated by the pope, namely that man cannot presume to know how the world is made, and must not restrict God's "omnipotence" (from the Latin meaning all-powerful).

Galileo was bold enough to present his ideas in the form of a dialog between two characters who voice arguments for and against both views of the cosmos. In it, Simplicio is given the job of speaking for the traditional, Earth-centered point of view. This is not a very subtle choice of name, and Salviati, the character representing the Copernican view, has no difficulty in showing Simplicio's arguments to be absurdly weak.

Moreover, Salviati demonstrates that all planets move around the Sun. Galileo believed too strongly in the Copernican argument to try to appear impartial.

The *Dialog on the Two Chief Systems of the World* was published in 1632 to wide acclaim, but the Church reacted strongly to it. Galileo was summoned to Rome to defend himself against charges of suspicion of heresy. At first he delayed his journey, using chronic ill-health as a reason. But a threat from Rome that if he did not come freely he would be "transported in chains" hastened his departure.

Galileo on Trial

Galileo had hoped for support from the grand duke of Tuscany. But when the time came Ferdinand II was strongly urged to put aside all respect and affection he might have for Galileo and show himself ready to shield Catholicism from danger. He heeded the warning and did not intervene on Galileo's behalf.

On June 21, 1633, Galileo stood before the Roman Inquisition. He was forced to recant—formally withdraw—his views, publicly affirming that he held "as very true and undoubted, Ptolemy's opinion of the stability of the Earth and the motion of the Sun." According to legend he then whispered, "*Et pur su muove*" ("And yet it moves"), but it is highly unlikely that he would have made such a dangerous remark. Sentenced to life imprisonment, Galileo served the term under house arrest at Arcetri, near Florence.

Galileo's appearance before the Inquisition in 1633, as recreated in 1847 by Joseph von Gemälde (1796–1890). Theology, said Galileo, was the business of the Church; investigation of Nature was the business of a scientist, and in his writings he had only been accounting for his scientific observations.

which measured the time it took for an amount of sand to run through a glass, and candle clocks, which measured the time it took for a wick to burn down marked-off divisions on a candle, were not precise enough, and the mechanical clocks in use at the time were extremely unreliable. Ingenious as ever, Galileo devised several systems of his own for measuring and and recording minute intervals of time (see box opposite).

BALLS DOWN A SLOPE

To aid his research, Galileo set out to find a way of slowing down as far as possible the rate at which objects fall. To do this, he rolled a smooth ball down a polished grooved slope, noting the time it took to reach the bottom. When he had repeated the experiment several times to test the accuracy of his results, he next rolled the ball one-quarter the length of the groove. His measurements showed that the ball took half the time to travel this distance as it took to travel the full length of the groove. "Next we tried other distances," he recorded, "comparing the time for the whole length with that for the half, or...for two-thirds."

By repeating these experiments as many as a hundred times, he was able to demonstrate that the distance covered by the falling object was proportional to the square of the time taken. This means that if, for example, an object falls 6 feet (1.83 meters) in one second, then in two seconds it will fall 2^2 (four times) as far; that is, it will fall a distance of 4 x 6 feet, or 24 feet (7.32 m). In three seconds it will fall 3^2 (nine times) as far, so it will fall 9 x 6 feet, or 54 feet (16.47 m), and so on. In other words, Galileo had discovered that, as an object falls, it accelerates at a consistent, or uniform, rate. He had therefore proved by his experiments what Isaac Newton (1642–1727) was later to describe in his second law of motion (see page 59).

FINAL YEARS

After his condemnation by the Inquisition, Galileo remained under house arrest at Arcetri until his death. Although he was unable to move about freely, he was allowed visitors. One intriguing meeting that took place toward the end of his life was with the English poet and philosopher John Milton (1608–1674), who made a point of visiting Galileo while on a tour of Italy in 1638–39. Galileo was also permitted to continue with his research, and to write. He began a new book on the sciences of motion and strength of materials. The last discoveries made with his telescope, recorded in 1637, noted the moon's diurnal (daily) and monthly movements (a side-to-side wobble).

Galileo was now suffering from high blood pressure, arthritis, and failing eyesight. This last ailment led him to the bitter comment that, though he had extended the visible universe a thousand times, for him it was now reduced "to a space no greater than that which is occupied by my own body." He died in 1642.

A 19th-century painting depicts Galileo rolling balls down a grooved slope to test the "law of fall." Though the picture has Pisan landmarks such as the Leaning Tower in the background, Galileo actually carried out these tests when living in Padua.

Keeping Time

In the early 17th century there were no timekeeping devices that could measure events accurately to the second, never mind to a fraction of a second. So in all his experiments, Galileo was forced to improvise methods of measuring time. Favorite methods included using his pulse rate, collecting drops of water, or humming tunes. Ingenious and effective as these were, Galileo was constantly searching for a more accurate method of recording time.

The story goes that, during his time studying medicine at the University of Pisa, the young Galileo attended a service in Pisa Cathedral. Gazing around him, he began watching a chandelier that was swinging in the wind. He noticed that the swings it made through the air were regular in pattern, and he started timing them using his pulse.

As he watched the chandelier he saw that sometimes it swung in small arcs, but at other times, such as when a sudden gust of wind blew through the cathedral, it swung in much longer arcs. However, Galileo observed that whatever length of arc the chandelier swung through, it took the same time to travel through it. "The Galileo chandelier" is still pointed out to people visiting Pisa Cathedral today.

In 1602 Galileo began experimenting for himself. He took two pendulums (weights suspended from a fixed point) of equal length and set them swinging through arcs of unequal length. Though one moved through a longer arc than the other, they both took the same time to swing through their arc, never going out of phase. He tried placing weights, equal and unequal, on the pendulums. Setting them swinging at the same time, he found they continued in harmony for 100, 500, or even 1000 oscillations (swings). Galileo realized that the pendulum could be used as a timing device. Galileo's son is supposed to have built the first working pendulum clock using this principle, though its design was perfected by Dutch scientist Christiaan Huygens (1629–1693) in 1659.

Galileo's experiments with pendulums pointed the way to the development of accurate timing devices.

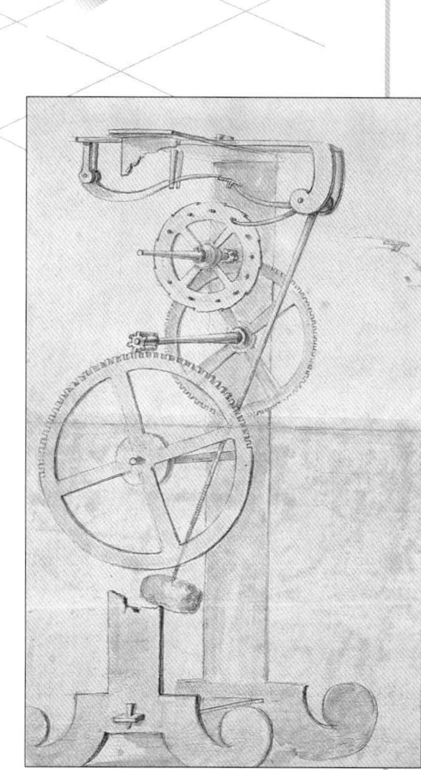

The swing of the pendulum (above) lasts for the same amount of time whatever the length of the arc it swings through. Galileo's pupil Vincenzio Viviani (1622–1703) drew a sketch of Galileo's design for an clock (right) based on this principle. The mechanism uses a toothed "escape" wheel and an "anchor" to regulate the movement of the pendulum. The clock was not actually built until the 19th century.

Science and Politics

In his role as chief mathematician at the court of the grand duke of Tuscany, Cosimo de' Medici II, Galileo taught the future grand duke, Ferdinand II, and his brother Leopold. The Medicis were a powerful Italian family who controlled Florence and Tuscany for most of the period from 1434 to 1737. Originally a Tuscan peasant family, they moved to Florence in the 12th century and accumulated vast wealth as bankers and merchants. Skillful operators, they bribed their way to power in politics and the Church, and also used their money to enrich the city of Florence. In this role, they were generous supporters of both the arts and the sciences. As early as 1442, Cosimo de' Medici, the founder of the family's political fortunes, had set up the Platonic Academy in Florence, in imitation of the original Academy in Athens (see box page 7).

In 1657, after Galileo's death, Ferdinand and Leopold de' Medici gave funds for a new Academy of Experiment in Florence. Leading figures in the Academy were two scientists who had also been taught by Galileo: Vincenzio Viviani and Evangelista Torricelli (1608–1647). Torricelli, who succeeded Galileo as mathematician to the grand duke, established the basic principles of hydromechanics (the mechanical properties of fluids) and invented the mercury barometer.

Unlike most scientific academies of the period, the Academy of Experiment encouraged members to carry out research and provided equipment for experiments. But it was closed down in 1667, probably for political reasons, when Leopold became a cardinal.

Powerful patron: a portrait of Cosimo II, the grand duke of Tuscany, by the Italian artist Pontormo (1494–1552)..

In Rome, the Academy of the Lynxes had been founded in 1603. It owed its name to the fact that the lynx was traditionally the most keen-sighted of animals. When news of Galileo's telescopic discoveries spread it turned its attention to astronomy and elected Galileo a member. Although the Academy helped pay for his publications, it withdrew its support when Copernicanism came under attack from the Church. The Academy of the Lynxes ceased its activities in the 1620s but was later revived and still exists today.

GALILEO'S LEGACY

Galileo saw the world as a Book of Nature that could be looked upon by everyone, but he emphasized that "it cannot be understood unless one first learns to comprehend the language and interpret the characters in which it is written." He went on to say that, "It is written in the language of mathematics, and its characters are triangles, circles, and other geometrical figures...without these one is wandering about in a dark labyrinth." Mathematics was central to all Galileo's work, and his contribution to mechanics, the mathematical study of bodies in motion, was outstanding. He undoubtedly paved the way for Newton's work on motion later in the 17th century.

Many of Galileo's predictions were confirmed by scientists working after him. He prophesied, for example, that more planets would eventually be observed beyond Saturn, which is the farthest planet clearly visible from Earth with just the naked eye. He was correct in this belief: Uranus was discovered in 1781, Neptune in 1846, and Pluto as

This imaginative 19th-century painting by Tito Lessi (1858–1917) recreates the scene when John Milton, the English poet and philosopher, visited Galileo at Arcetri. Milton, on the left, listens eagerly as the elderly scientist expounds his theories.

The house at Arcetri, near Florence, where Galileo lived under house arrest. From here he continued his work, writing *Discourses upon Two New Sciences,* which was published in Leiden, Holland, in 1638.

recently as 1930. Galileo was also right to state that light travels at a very high speed and to predict that scientists would eventually find the means of measuring and recording its rate of passage.

Lifting the Condemnation

Even after his death in 1642, the Church authorities continued their vendetta against Galileo. He was not permitted a public funeral, or allowed to be buried next to his father and other family members. Although a number of memorials to Galileo can be seen today in Florence, they were all built later.

In 1559 the papal authorities introduced an *Index of Prohibited Books*. This consisted of a list of books that the Church considered unsuitable to be read by ordinary men and women because of the views they expressed. Any reader of one of these forbidden books faced possible excommunication (exclusion from the Church). The works of Copernicus were placed on the index in 1616, and in 1664 Galileo's *Dialog on the Two Chief Systems of the World* was added to it. Galileo's work was not removed from the list until 1835, and it was only in 1992, 350 years after Galileo's death, that Pope John Paul II (1920–) formally announced that the Inquisition had been wrong to condemn Galileo, and that Galileo had been right to claim that the Bible could not always be taken literally.

◀ ***see also*** Copernicus VOL 1:14
▶ ***see also*** Kepler VOL 1:38 Newton VOL 1:54

GALILEO: Life and Times

SCIENTIFIC BACKGROUND

Before 1560

Polish astronomer Nicolaus Copernicus (1473–1543) proposes a new heliocentric (Sun-centered) view of the universe

Italian mathematician Giovanni Battista Benedetti (1530–1590) shows that the speed of falling bodies is not related to their weight

1560

1570

c. 1570 Italian physicist Giambattista della Porta (1535–1615) describes a camera obscura (pinhole camera) fitted with a lens

1575 The Academy of Mathematical Sciences is founded in Madrid, Spain

1580

1581 Galileo watches a chandelier swinging in Pisa cathedral and notes that the time of swing remains the same whatever the length of arc it swings through

1588 Danish astronomer Tycho Brahe (1546–1601) rejects Aristotle's idea that the planets are held by transparent crystalline spheres

1589 German astronomer Johannes Kepler (1571–1630) is converted to Copernicus's Sun-centered view of the universe

1590

1590 In his book, *Motion*, Galileo rejects Aristotle's ideas about falling bodies

c. 1593 Galileo invents an early form of thermometer

1597 Writing to Kepler, Galileo admits he accepts Copernicus's view of the universe

POLITICAL AND CULTURAL BACKGROUND

1556 Charles V (1500–1558), the most powerful monarch in Europe, abdicates as Holy Roman emperor and king of Spain

1559 The wars between France and Spain for supremacy in Italy are ended with the Treaty of Cateau-Cambrésis

1566 The Dutch revolt against Spanish imperial rule, beginning a long war for independence

1569 Cosimo de' Medici I (1519–1574), known as Cosimo the Great, becomes Grand Duke of Tuscany

1572 Thousands of French Protestants are slaughtered following the massacre of St. Bartholomew in Paris

1577–80 English navigator Francis Drake (1540–1596) sails around the world in the *Golden Hind*

1582 Pope Gregory XIII (1502–1585) reforms the calendar; October 5 now becomes October 15

1588 Philip II of Spain (1527–1598) sends an invasion fleet, the Armada, against England, but it is scattered by storms in the Channel

1593 Henry of Navarre (1553–1610), a Protestant, converts to Catholicism on becoming King Henry IV of France

c. 1595 English playwright William Shakespeare (1564–1616) completes his tragic romance, *Romeo and Juliet*

1598 The Edict of Nantes establishes religious toleration in France

1600

1600 Dutch lens grinders construct the first refracting and compound telescopes

1604 Galileo calculates that the distance covered by objects in free fall is proportional to the square of the time taken

1609 Galileo constructs a refracting telescope

1609 Kepler advances his theories that planets revolve round the Sun in elliptical orbits, and that planets move faster nearer the Sun and more slowly when farther away

Bruno (1548–1600), supporter of the Copernican system of the universe, is burned at the stake after a seven-year trial held by the Inquisition

1603 James VI of Scotland (1566–1625) becomes King James I of England on the death of Queen Elizabeth I (1533–1603)

1607 Jamestown, Virginia, is established; the colonists elect Captain John Smith (1580–1631) as their leader

1610

1610 Using his telescope, Galileo observes Jupiter's moons and individual stars in the Milky Way; he reports his findings in *The Starry Messenger*

1613 Galileo publishes *Letters on Sunspots*, which supports the Copernican system

1616 The Roman Catholic Church condemns the heliocentric view of the cosmos as heretical

1618 Galileo becomes involved in a controversy about comets

1615 Spanish author Miguel de Cervantes (1547–1616) completes his masterpiece, *Don Quixote*

1617 Pocahontas (1595–1617), the Native American princess who intervened twice to save the life of Captain John Smith, dies in a smallpox epidemic in England

1620

1623 *The Assayer*, a work by Galileo concerning scientific methodology, is published in Rome. He begins work on a major study of cosmology

1620 The Pilgrim Fathers land in North America

1621 After the death of Cosimo II, Grand Duke of Tuscany, his young son, Ferdinand II (1610–1670), succeeds to the title

1623 Maffeo Barberini (1568–1644) is elected Pope Urban VIII; he supports the Inquisition's condemnation of Galileo but later reduces his sentence to house arrest

1624 The cardinal-duke of Richelieu (1585–1642), favorite of King Louis XIII (1601–1643), takes sole control of affairs in France

1630

1632 Galileo publishes his highly contentious *Dialog on the Two Chief Systems of the World*

1633 Galileo is sentenced by the Roman Inquisition to spend the rest of his life under house arrest

1638 Galileo's *Dialogs upon Two New Sciences* discusses laws of motion and friction

1632 Sweden's King Gustavus II Adolphus (1594–1632), leader of the Protestant army in the Thirty Years War (1618–48) is killed, aged 38, at the Battle of Lutzen

1637 French philosopher and mathematician René Descartes (1596–1650) completes the first of his major philosophical works, the *Discourse on Method*

1640

After 1640

1643 Galileo's pupil Evangelista Torricelli (1608–1647) invents a mercury barometer

1657 Christiaan Huygens (1629–1693) builds the first pendulum clock

1657 Accademia del Cimento, the first experimental scientific research center since ancient times, is established in Rome

1687 Building on Galileo's work, English scientist Isaac Newton (1642–1727) establishes his laws of motion

1643 Louis XIV (1638–1715) is crowned King of France; he will become known as the Sun King

Johannes Kepler

1571–1630

"I vowed to God I would make public in print this wonderful example of his Wisdom."

Johannes Kepler
The Mystery of the Universe
(1597)

THE GERMAN ASTRONOMER JOHANNES KEPLER WAS THE FIRST PERSON TO OBSERVE THAT THE EARTH ORBITS THE SUN ON AN ELLIPTICAL (OVAL) PATH. HIS IDEAS REVOLUTIONIZED THE STUDY OF ASTRONOMY AND LED TO THE DISCOVERY OF THREE LAWS OF PLANETARY MOTION THAT ARE STILL IN USE TODAY.

JOHANNES KEPLER WAS BORN IN WEILDERSTADT, Württemberg, south Germany, the premature son of Lutheran parents. In infancy he almost died of smallpox and it seems that he was plagued by ill-health from then on. His letters and manuscripts frequently contain references to his various illnesses. At the age of 14 he suffered continually from "skin ailments, severe sores, scabs of chronic putrid wounds in my feet.... On the middle finger of my right hand I had a worm, on the left a huge sore." Later he "began to suffer from headaches, mange," and mental disturbances. Whether he was constantly sick or simply imagined the illnesses, it is difficult to say.

In 1589 Kepler entered Tübingen University, where he studied theology with the intention of becoming a Lutheran priest. The course included lectures in mathematics and astronomy; it was there that Kepler was introduced to Copernicus's theory that the Earth and all the planets orbit around the Sun (see page 17). In 1594 Kepler was invited to teach mathematics at a high school in Gräz, Austria. He accepted the offer and abandoned theology in order to devote himself to the study of astronomy and related subjects.

In Search of a Divine Pattern

As his work got underway, Kepler quickly became convinced that he had discovered the ultimate secret, that he was the first person to know exactly what was in God's mind when he constructed the solar system. Unfortunately, Kepler's first "great discovery" was based upon a false assumption, namely that there are only six planets orbiting the Sun: Mercury, Venus, Earth, Mars, Jupiter, and Saturn. The three other planets in our solar system were still unknown in Kepler's day: Uranus was discovered nearly 200 years later in 1781, Neptune in 1846, and Pluto in 1930.

KEY DATES	
1571	Born in Weilderstadt, Württemberg, Germany on December 27
1589	Enters the University of Tübingen
1594	Appointed professor of mathematics at Gräz
1597	Marries first wife, Barbara
1601	On Tycho Brahe's death, appointed imperial mathematician at the court of Emperor Rudolf II in Prague
1609	Publishes first two laws of planetary motion in *The New Astronomy*
1611	Death of wife and son
1612	Moves to Linz as imperial mathematician
1613	Marries second wife
1619	Publishes third law of planetary motion in *Harmony of the World*
1620	Mother tried for witchcraft
1628	Appointed mathematician to Albrecht Wallenstein, commander-in-chief of the imperial forces
1630	Dies in Regensburg, Germany, on November 15

A 19th-century etching shows Johannes Kepler (standing) in consultation with the Danish imperial astronomer Tycho Brahe (1546–1601) in the emperor's observatory in Prague. Brahe did not share many of Kepler's astronomical theories, but he recognized Kepler's extraordinary talent for mathematics and astronomy.

Kepler began to ask himself why there were six planets, and why they were located where they were in the heavens. Kepler was sure that God must have had a divine purpose in the way he ordered the universe, and thought this must be based on some kind of mathematical principle. He reasoned that, as planets moved in three-dimensional space, their orbits might be related to three-dimensional objects such as spheres and cubes. It was at this point that Kepler had an interesting idea.

Between the six orbits of the six planets there are five spaces. Kepler was familiar with the ancient geometry of the Greek mathematician Euclid (active c. 300 BC), and the number five brought to his mind the number of "regular solids" as described by Euclid. A regular solid is one with identical faces: the cube is an example, with six square faces. Only five such solids can exist (see above right); not even God could have created a sixth. All the regular solids can be "inscribed" within a sphere (their points touch the inside of the sphere), and can themselves inscribe a sphere (their planes are in contact with the sphere's outer side).

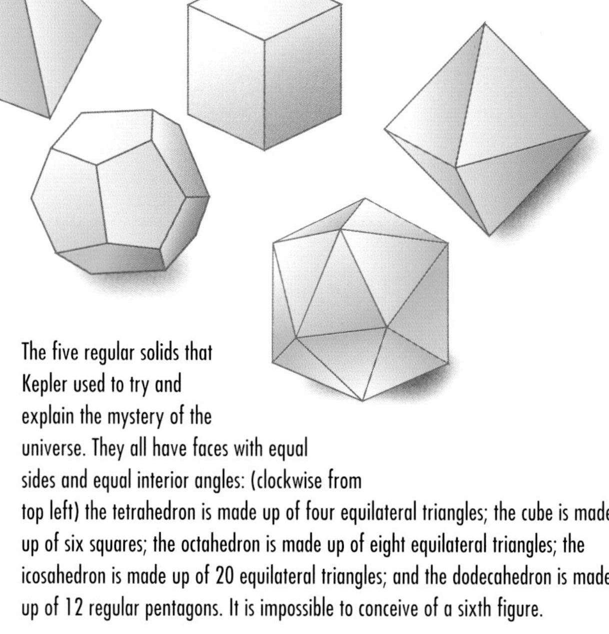

The five regular solids that Kepler used to try and explain the mystery of the universe. They all have faces with equal sides and equal interior angles: (clockwise from top left) the tetrahedron is made up of four equilateral triangles; the cube is made up of six squares; the octahedron is made up of eight equilateral triangles; the icosahedron is made up of 20 equilateral triangles; and the dodecahedron is made up of 12 regular pentagons. It is impossible to conceive of a sixth figure.

Kepler suggested that the cosmos might be made up of six spheres (the orbits of the planets), within which the dimensions of the five regular solids could be inscribed. This would explain why God had created just six planets—there simply was no room for a seventh. Kepler also thought that the ratios between planetary distances might be mirrored by the dimensions of the regular solids, and that this would be a way to explain the distribution of the planets in space. He produced a model of his planetary system showing how he imagined it would work (see opposite).

Kepler checked his work by comparing the distances between planets derived from his own model with those established by Copernicus, and found them to fit reasonably well. Where they differed by a large amount, Kepler thought that Copernicus must have made mistakes. In 1596 he published a book, *A Precursor to Cosmographical Treatises Containing the Mystery of the Universe,* in which he revealed his great discovery and reported how: "The intense pleasure I have received from this discovery can never be told in words.... I vowed to God that I would make public in print this wonderful example of his Wisdom." We now know that this

Tycho Brahe
1546–1601

Brahe was born into a Danish aristocratic family. At the age of 13 he was sent to the University of Copenhagen to study for a career in politics; while there he observed a partial eclipse of the Sun and decided to become an astronomer. As a young man he lost part of his nose in a duel, and replaced the missing piece with a metal substitute.

In 1576 Brahe was given the island of Hven as a gift by the Danish king, Frederick II (1534–1588), and there he set up the first major observatory in Christian Europe; in effect, he ruled the island as his own kingdom. The observatory was stocked with the most accurate astronomical instruments available, and during the 1580s and 1590s Brahe collected a mass of data about the planets. These would later serve as an important source for Kepler's calculations.

Brahe made his reputation in 1572 with the observation and description of a new star—since called Tycho's star—and of the comet of 1577. He noticed that the comet was moving through the areas between the planets, and realized that the planets could not be carried in transparent spheres, as astronomers had previously thought.

In 1597 Brahe fell out of favor with Frederick II's successor. He moved to the imperial court at Prague, where he served as mathematician to Emperor Rudolf II (1552–1612). In 1600 Kepler became his assistant, gaining access to Brahe's data after his death in 1601.

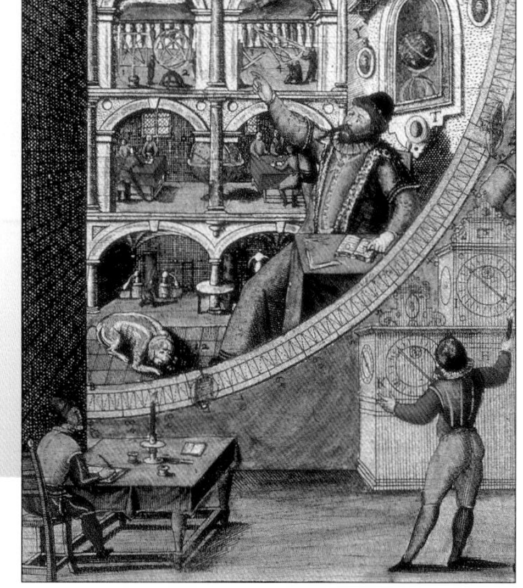

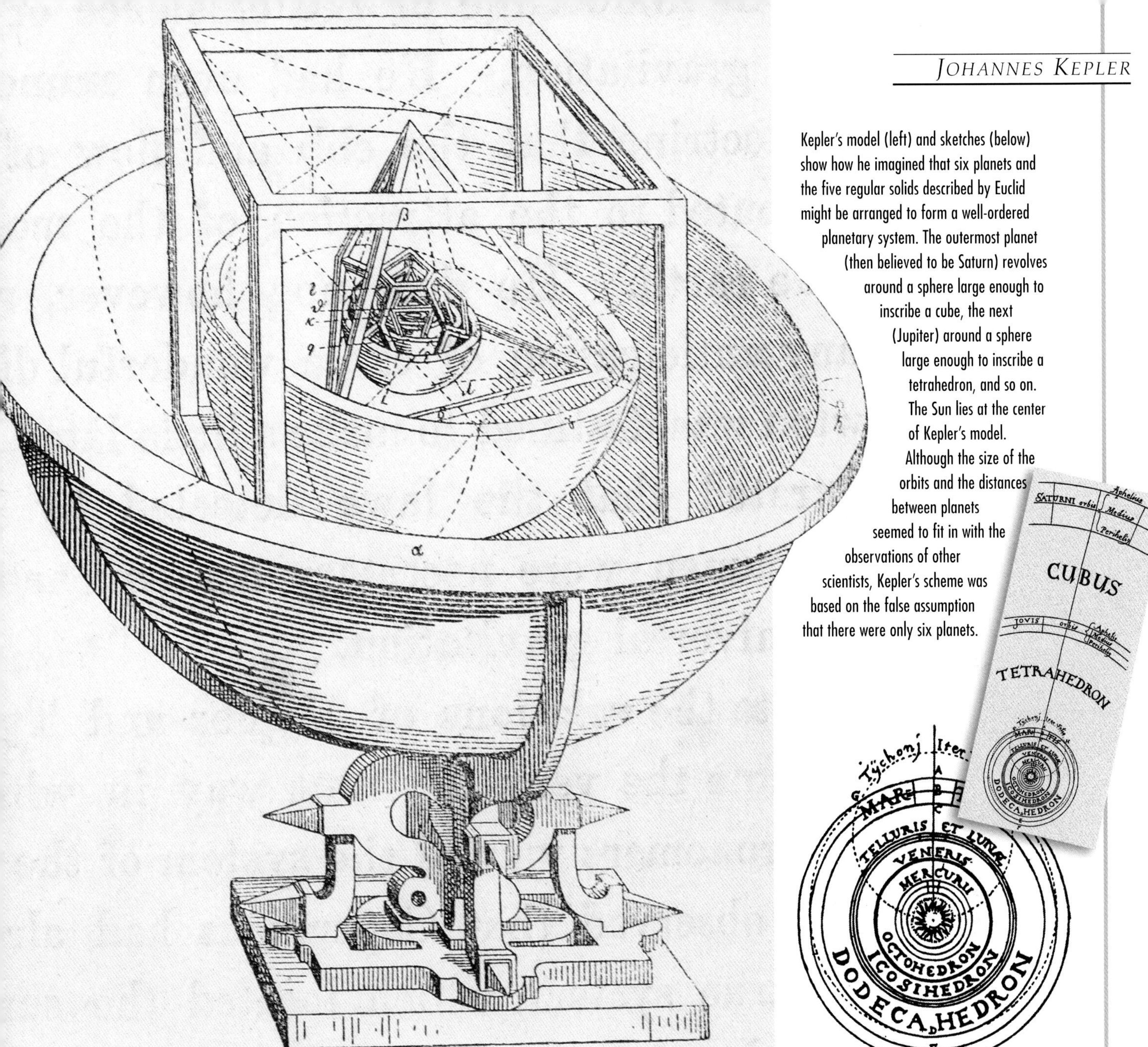

Kepler's model (left) and sketches (below) show how he imagined that six planets and the five regular solids described by Euclid might be arranged to form a well-ordered planetary system. The outermost planet (then believed to be Saturn) revolves around a sphere large enough to inscribe a cube, the next (Jupiter) around a sphere large enough to inscribe a tetrahedron, and so on. The Sun lies at the center of Kepler's model. Although the size of the orbits and the distances between planets seemed to fit in with the observations of other scientists, Kepler's scheme was based on the false assumption that there were only six planets.

theory was false, but the scientists of the day were extremely impressed. At the very least, Kepler had proved himself to be a daring thinker.

Imperial Approval

Kepler sent a copy of *The Mystery of the Universe* to several scientists, among them Tycho Brahe, the Danish astronomer and mathematician. Brahe did not share Kepler's belief in the Copernican planetary system; Brahe himself had produced a new scheme of the cosmos that retained the Earth at the center of the universe, even if he agreed that the other planets might orbit the Sun. However, he recognized Kepler's talents. In 1599 Brahe became imperial mathematician to the Emperor Rudolf II, and invited Kepler to become his assistant at his observatory near Prague in Bohemia (now the Czech Republic). Kepler joined him in 1600, and when Brahe died the following year, Kepler was appointed his successor as imperial mathematician. Kepler now had access to the whole of Brahe's carefully recorded observations compiled over many years of sky-watching.

Stars and Mars

In October 1604, Kepler saw and described a supernova, a star that brightens temporarily by several hundred billion times its original brightness. The

Kepler's Three Laws

Kepler discovered three important laws concerning planetary motion. The first two were announced in 1609. The first states that planets move around the Sun in elliptical orbits, with the Sun at one focus, or one end, of the ellipse (see below). This was a ground-breaking suggestion because until this time all astronomers, including Kepler himself, had always assumed that planets moved in circular orbits.

For this reason it has often been suggested that Kepler's first law marks the true dividing line between ancient and modern astronomy. The law took several years to formulate, and was based on Kepler's efforts to understand the orbit of Mars. Kepler found that, however hard he tried, he could not fit the orbit of Mars to any circular path.

Eventually he decided that, if its path was not circular, Mars must travel in an elliptical orbit. Detailed calculations on this point led Kepler to formulate his second law. This states that an imaginary line connecting the Sun and a planet will sweep out equal areas of space in equal times as the planet follows its elliptical course around the Sun. Put more simply, this meant that the speed of a planet's orbit slowed as it got farther from the Sun, and speeded up as it got nearer to it.

Kepler's second law shattered the assumption held by astronomers ever since the time of the ancient Greeks that the planets always move at a steady (uniform) speed. It implied that the Sun must somehow control their speed, though Kepler was unable to explain how that might be. His ideas, however, prepared the way for the work of English scientist Isaac Newton (1642– 1727) later in the 17th century, when he came to the realization that the same force that causes objects to fall to the ground (namely gravity) also keeps the planets traveling in their orbits (see page 61).

Kepler's observation of Mars showed that its path around the Sun was elliptical, and that the Sun is positioned at one end (focus) of the ellipse. The lines and positions on the ellipse show wedges of equal areas and segments representing equal times on the orbit. It is clear that, when it is nearer the Sun, Mars covers a greater distance along the ellipse in a shorter time, because it moves faster.

Nearly a decade passed before Kepler published his third law in *Harmony of the World* (1619). This stated that the ratio between the time one planet takes to orbit the Sun and its average distance from the Sun is the same for all the other planets. Astronomers could see from their observations of the skies how long it takes a planet to orbit the Sun. So, once they were able to calculate the distance from the Sun of

Danish astronomer Tycho Brahe's very detailed astronomical observations were invaluable in helping Kepler develop his revolutionary laws of planetary motion. However, Brahe's own scheme for the cosmos, illustrated here, still held to the view that the planets revolve around the Earth.

just one planet, they could work out exactly what the ratio described in Kepler's third law was. Using this ratio, it would then become possible to work out how far away from the Sun all the other planets were.

supernova was slightly dimmer than Venus, the brightest of the planets, and stayed in the sky for more than a year. Its appearance disproved the ancient idea that there was a region of fixed stars in the heavens that never changed.

As he continued his observation of the skies, Kepler became more and more convinced that the movement of Mars did not fit with Brahe's Earth-centered planetary system, or with Copernicus's idea that all the planets revolved around the Sun in a perfectly circular path. In 1609 he published another book, *The New Astronomy*, in which he set out his revolutionary ideas. Not only could he prove that Mars orbited the Sun, but also that it did so on an elliptical path. He had cast aside his own earlier model of a planetary system based on orbits and solids, and now went on to draw up the three laws of planetary motion for which he is best remembered (see box).

KEPLER'S WINE BARRELS

Kepler did not confine his scientific investigations to astronomy. One example of his endlessly inquiring mind is provided by the story of the wine barrels. He noted that the wine barrels delivered to his household were of various shapes and began to wonder how much wine was in each barrel; the merchants seemed to use very rough methods to calculate how much they held. Kepler realized that if he imagined a barrel to be made up of hundreds of very thin, round sections, he could calculate the area of each circular piece individually, and then add them together again. This would give him a much more accurate picture of what volume of wine each barrel held. Kepler's method anticipated the principle behind the calculus, the branch of mathematical reckoning developed independently by Isaac Newton (see page 66) and the German mathematician Gottfried Leibniz (1646–1716).

PUBLIC AND PRIVATE TURMOILS

Europe was passing through a period of intense political and religious tension, and the city of Prague was right at the center of the troubles. In 1618, the violent unrest forced Kepler to leave the city. It was the beginning of the Thirty Years War, a bitter conflict between Catholics and Protestants in Europe that lasted until 1648.

Omens and Portents

Today we regard astrology, the belief that the movements of the planets, Sun and Moon can influence the outcome of events on Earth, as having no scientific basis. The distinction was not so great in Kepler's day.

Astrology originated in Mesapotamia in about the 3rd millennium BC. By about 400 BC astronomers had adopted the 12 houses of the zodiac, using them to construct personal predictions. The Greeks absorbed Near Eastern beliefs that the stars and planets were divine and directly linked to happenings on Earth, and by 300 BC Greek astronomers were drawing up "horoscopes" (from the Greek for "ascendant birth sign").

The Romans adopted and developed the ideas of astrologers, and they also influenced early Christian thinkers. Despite the discovery that the Sun, not the Earth, lay at the center of the universe, astrological prediction remained enormously popular in 17th-century Europe, and astronomers were kept busy producing almanacs. These listed dates of the movable Christian festivals, such as Easter, and noted significant astronomical events such as eclipses or new moons; but they might also include predictions based on planetary activity.

Although Kepler rejected the idea that the stars guide the lives of human beings, and later called astrology "the foolish little daughter of astronomy," his imperial patron, Rudolf II, was a keen student of astrology. Kepler drew up an almanac in which he correctly predicted very cold weather and a Turkish invasion, and thereafter his astrological skills were much in demand in Prague, bringing him financial reward. In 1628 Kepler was appointed mathematician to the imperial general Albrecht Wallenstein (1583–1634). He composed a horoscope for the general predicting "horrible disorder" before March 1634. Wallenstein was assassinated in February that year.

Some of Kepler's astrological writings preserved in Regensburg (below), and a 1608 horoscope (inset) prepared by Kepler.

Kepler's first wife Barbara—described by him as "simple of mind and fat of body"—had died in 1611. Resettled in Linz in Austria, Kepler began to look for someone else to marry. He seems to have considered as many as 11 women before settling on one who was "modest, thrifty, diligent and loved her stepchildren."

More worrying personal concerns faced Kepler in 1620 when his mother was arrested as a suspected witch. This was a serious matter: in Kepler's home town alone, 38 witches were burned between 1615 and 1629. His mother was held in prison for over a year, but set free after a trial.

Kepler's *Rudolphine Tables* (named for his former patron, Rudolf II) were published in 1627. They were tables of planetary motion based on Brahe's observations. The following year Kepler and his family were forced to leave Linz; they settled in Silesia (now southwest Poland). Two years later, while traveling alone to Austria, Kepler fell ill and died in Regensburg, southern Germany.

◀ **see also** Copernicus VOL 1:14 Galileo VOL 1:24
▶ **see also** Newton VOL 1:54

KEPLER: Life and Times

SCIENTIFIC BACKGROUND

Before 1550

Greek mathematician Euclid (active c. 300 BC) describes the five regular solids

Greek philosopher Aristotle (384–322 BC) concludes that stars and planets are supported by crystal spheres

Polish astronomer Nicolaus Copernicus (1473–1543) publishes *On the Revolutions of the Celestial Spheres* (1543), in which he puts forward his theory that the Earth travels around the Sun

1550

1560

1560 In Naples Italian physicist Giambattista della Porta (1543–1615) founds Europe's first scientific society, the Academy of the Secrets of Nature

1570

1572 Danish astronomer Tycho Brahe (1546–1601) observes a new star, disproving Aristotle's theory that nothing in heaven changes

1577 Brahe observes the great comet and calculates that it must be at least four times farther from Earth than the Moon

1580

1588 Brahe rejects the idea that all planets are supported by crystal spheres

1589 Kepler is converted to Copernicus's Sun-centered view of the universe

1590

1596 In *The Mystery of the Universe*, Kepler reveals his belief that the cosmos is made up of six planets

1600

1609 Kepler publishes *The New Astronomy*, in which he suggests that planets revolve around the Sun in an elliptical orbit

1610

1610 Italian mathematician and astronomer Galileo Galilei (1564–1642) observes the moons of Jupiter, the stars of the Milky Way, and the phases of Venus

1614 Scottish mathematician John Napier (1550–1617) introduces logarithms for calculating numbers

1619 Kepler's third law establishes a ratio between length of planetary orbit and distance from the Sun

1620

1627 Kepler publishes the *Rudolphine Tables*, detailing planetary motion

1630

After 1630

1660s English scientist Isaac Newton (1642–1727) develops the calculus; Kepler had used a similar system to work out the volume of wine in a barrel

1687 Newton demonstrates his picture of the universe, based on the ideas of Kepler, Copernicus, and Galileo, in his *Principia*

POLITICAL AND CULTURAL BACKGROUND

1560 Scottish Protestant reformer John Knox (c. 1513–1572) founds the Presbyterian Church

1568 Italian artist and art historian Giorgio Vasari (1511–1574) publishes his great work, *The Lives of the Most Eminent Painters, Sculptors, and Architects*

1569 Flemish mapmaker Gerardus Mercator (1512–1594) devises the Mercator map projection, which has been used on nautical charts ever since

1573 In China 10-year-old Wan Li (1563–1620) assumes the imperial throne; he will reign for 46 years

1576 Rudolf II (1557–1612) becomes Holy Roman emperor. He makes his capital at Prague a center for writers and scholars, and is a supporter of Brahe and Kepler

1587 The legend of Dr Faustus, the astrologer who sold his soul to the devil in exhange for knowledge and power, is published anonymously in Germany and quickly becomes popular throughout Europe

1600 Italian philosopher Giordano Bruno (1548–1600), a champion of Copernicus's Sun-centered theory of the universe, is burned at the stake for his unorthodox religious views

1613 The Romanov dynasty assumes power in Russia; Romanovs will rule there for almost 300 years

1614 Virginia colony widower John Rolfe (1585–1622) marries American Indian Pocahontas (1595–1617)

1618 A rebellion against imperial rule in Prague (the Defenestration of Prague) begins the Thirty Years War, which lasts until 1648

1620 The English Pilgrim Fathers land at Massachusetts to found the Plymouth colony

1624 Albrecht von Wallenstein (1583–1634) becomes commander of the Catholic imperial army in the Thirty Years War

1633 The Roman Catholic Inquisition forces Galileo to recant his Copernican view of the universe

WILLIAM HARVEY

1578–1657

"...There is but one road to science, that...in which we proceed from things more known to less known...; the comprehension of universals by the understanding is based upon the perception of individual things by the senses."

SIR WILLIAM HARVEY
Anatomical Exercises on the Generation of Animals
(1651)

ENGLISH PHYSICIAN WILLIAM HARVEY HELPED LAY THE FOUNDATIONS OF MODERN MEDICINE AND PHYSIOLOGY (HOW LIVING ORGANISMS FUNCTION). HIS WORK, BASED ON DIRECT OBSERVATION AND CAREFUL EXPERIMENT, CHALLENGED LONG-HELD MEDICAL OPINIONS AND PROVIDED CONCLUSIVE EVIDENCE THAT THE HEART ACTS AS A PUMP, CIRCULATING BLOOD THROUGHOUT THE BODY.

WILLIAM HARVEY WAS BORN ON APRIL 1, 1578, in Folkestone, Kent, southeast England, into a prosperous farming family. He was the eldest of seven sons, the only one to embark on a scientific career. His brothers became successful merchants in London.

When he was 10 years old William began his education at King's School, Canterbury. He won a scholarship to study medicine at Cambridge University, and entered Caius College in 1593, graduating in 1597. At the time, this was considered the best place in England to study medicine.

In 1599 William moved to the medical school at Padua University in Italy to study under the Italian Girolamo Fabrizio (1537–1619). He is usually known as Fabricius, the short Latin version of his name, or, in full, as Hieronymus Fabricius ab Aquapendente (Aquapendente is the town in northern Italy where he was born). Fabricius specialized in the study of anatomy (the physical structure of animals and plants), and embryology (the study of embryos). The great astronomer and mathematician Galileo Galilei (1564–1642) was then professor of mathematics at Padua; his method of working out scientific laws from observation and experiment was attracting great attention. Harvey was strongly influenced by it, but probably never met Galileo.

A PRACTICING PHYSICIAN

Harvey obtained his medical degree in 1602 and returned to England, setting up in practice as a physician in London. He was very successful. His patients included Francis Bacon (1561–1626), lord chancellor of England, and the earl of Arundel, who became ambassador to the court of Ferdinand II (1578–1637), the Holy Roman emperor. Harvey

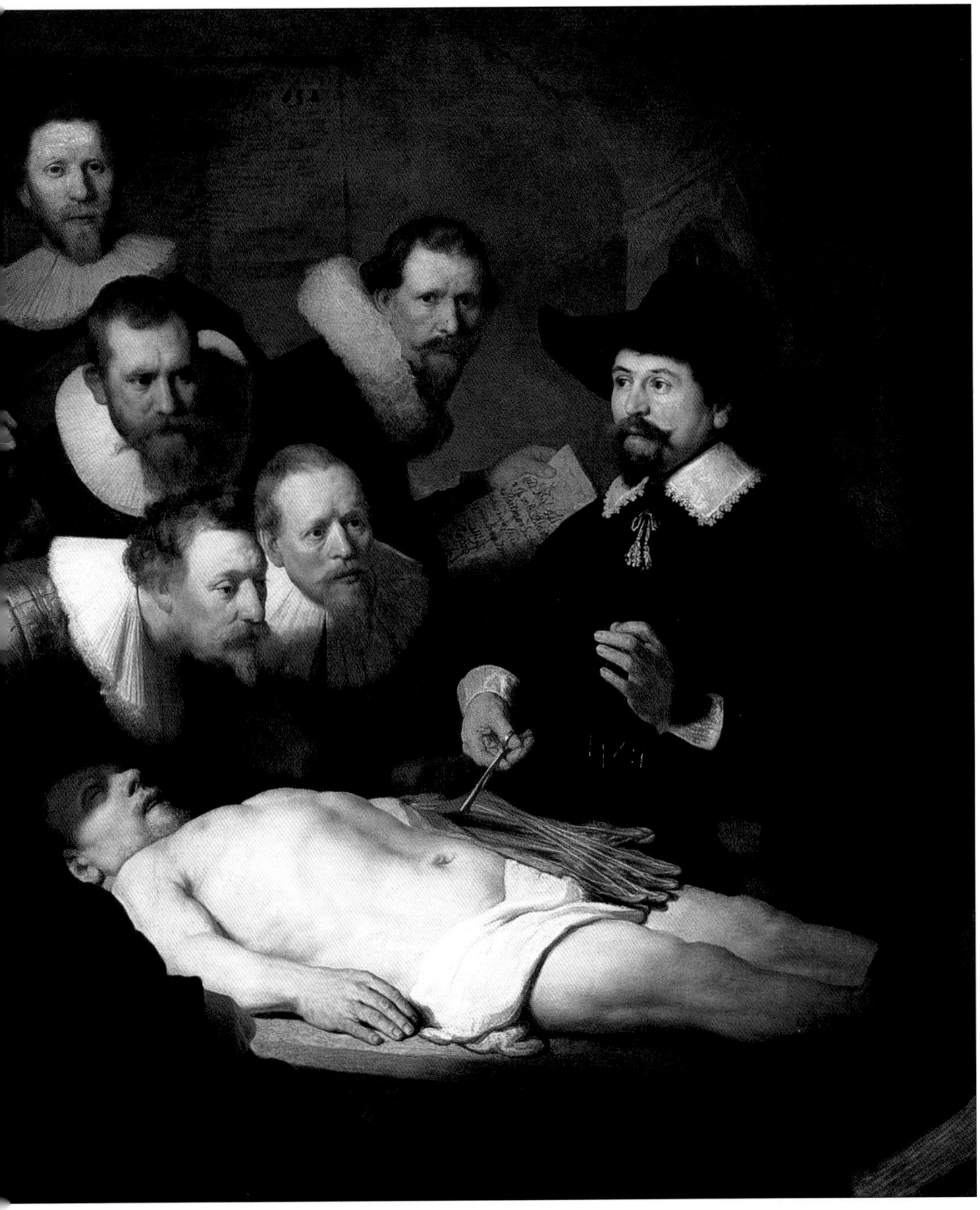

KEY DATES

1578	Born on April 1
1588	Starts at King's School, Canterbury
1593	Enters Caius College, Cambridge University
1597	Graduates from Cambridge
1599	Enters the medical school of the University of Padua
1602	Graduates from Padua
1604	Marries Elizabeth Browne
1607	Elected a fellow of the Royal College of Physicians
1609	Appointed physician to St. Bartholomew's Hospital
1615–43	Professor at St. Bartholomew's
1616	Begins course of lectures on the heart and blood
1618–49	Serves as physician to King James I and then King Charles I
1628	Publishes work on circulation in *Anatomical Exercises on the Motion of the Heart and Blood in Animals*
1651	Publishes work on embryology
1657	Dies in London on June 3

This painting by the Dutch painter Rembrandt (1601–1669) depicts a doctor instructing his students in anatomy. He is demonstrating the workings of the wrist and fingers.

was married in 1604 to Elizabeth Browne, daughter of the physician to King James I (1566–1625).

In 1607 Harvey was elected a "fellow" (member) of the Royal College of Physicians, and in 1609 was appointed physician to St. Bartholomew's Hospital. From 1615 to 1643 he taught there as a professor. It was traditional for a physician to give lectures attended by surgeons, and in 1616 Harvey began a course of lectures in which he explained his ideas and discoveries about the heart and blood.

Two years later he succeeded his father-in-law as physician to King James I and, after James's

The courtyard of Padua University, where Harvey studied. Fabricius and Galileo both taught here.

death he remained at court as physician to his son, King Charles I (1600–1649). Between 1629 and 1632 Harvey traveled widely, especially in Italy.

In 1642 civil war broke out in England between the supporters of Parliament and the supporters of King Charles I. Harvey may have been present with the king's army at the battle of Edgehill in October 1642. Many of his research papers were destroyed when the Parliamentary forces ransacked the king's palace at Whitehall. He was made warden of Merton College, Oxford in 1645 but lost his post after King Charles's defeat.

In 1654 Harvey was elected president of the Royal College of Physicians, but at 76 he felt was too old, so declined the honor. He died in London, on June 3, 1657. John Aubrey (1626–97), the English diarist, described Harvey's appearance: "In person he was not tall, but of the lowest stature; round faced, olivaster [olive-ish] complexion, little eyes, round, very black, full of spirits; his hair black as a raven, but quite white 20 years before he died."

CIRCULATION OF THE BLOOD

Harvey was a highly successful physician, but his reputation rests on his medical research. While a student of Fabricius at Padua University, he became interested in the way blood moves through the body. He came to realize that none of the existing explanations was satisfactory.

The first person to describe blood in detail was Aristotle (384 BC–322 BC) in the 4th century BC. He thought that it was produced in the liver from food, carried through the body in the veins, and cooled by its passage through the brain. Two Greek physicians living in Alexandria, Egypt, developed these ideas. Herophilus (active 300–250 BC) noted that arteries (large tubular vessels in the body) pulsate, or vibrate regularly. However, he failed to connect this with the beat of the heart. Erasistratus (c. 304–c. 250 BC) thought that every organ in the body was served by veins, arteries, and nerves. He believed nerves were hollow and carried "nervous spirit," that arteries carried a kind of air called "animal spirit," and that veins carried blood from the heart.

Four centuries later Galen (c. 130–c. 201 AD), who lived at the height of the Roman empire, discovered that arteries carry blood, not air, and that nerves arise from the brain. He believed that blood

Healing the Sick

The origins of medical treatment in the West can be traced back to ancient Greece. Many philosophers attempted to understand the causes and treatment of disease. For example, the Greek physician Hippocrates of Cos (c. 460–377 or 359 BC) tried to link environment and disease in his book *Airs, Waters, and Places*. His name lives on in the Hippocratic code—a set of laws that describes the way doctors should behave toward their patients. Many doctors still swear the Hippocratic oath today.

In Arab countries, the medical texts of ancient Greece and Rome were preserved and studied by Islamic scholars, while they were all but forgotten in the West for many centuries. The practice of medicine flourished, and hospitals and medical schools were established in cities across the Muslim empire from Baghdad in modern Iraq to Córdoba in Spain.

During the Middle Ages in Europe, the care of the sick was a religious duty undertaken by monks and nuns in infirmaries, special buildings attached to monasteries and convents. In Paris, for example, the Hôtel Dieu (God's Hostel), an infirmary, was built onto the walls of Notre-Dâme cathedral. Such infirmaries often had herb gardens close by, where the medicinal plants needed for the treatment of patients were grown.

Early Medical Schools

The first medical school in Europe was founded in Salerno, Italy (a part of the world where Arab influences were strong), between the 9th and 11th centuries. The Hospital of the Holy Ghost in Montpellier, France, founded in 1145, became an important training center for doctors. But as universities spread through western Europe, those studying medicine were often persuaded to give up practical studies in favor of university professorships. Medical teaching was firmly rooted in the ideas of Greek writers such as Aristotle and Galen.

An anatomy class as shown in *On Anatomy* (1559) by Matteo Realdo Colombo (1516–1559), who was a professor at Padua University. The lecturer performed the dissection, while students consulted textbooks and took notes.

By the end of the 15th century medical care was increasingly moving outside the control of the Church. In nearly every village, wise women made use of age-old herbal remedies to treat the sick. But institutions such as the College of Physicians, founded in London in 1518, grew up to regulate the practice of medicine and to ensure that only qualified people could dispense advice. Physicians had to pass an examination before being allowed to practice. Harvey undertook three oral examinations before obtaining his full license in 1604.

Hospitals

For the most part, hospitals were simply refuges where the chronically sick and poor were looked after until they died; little or no attempt was made to cure people. But gradually the habit developed of teaching students by practical lectures at the patient's bedside. Little by little hospitals became more professional places.

In 1607 Harvey became a fellow of the College of Physicians, entitling him to work at St. Bartholomew's. Founded in 1123 as London's first hospital, it had to take in every sick person judged to be curable. Harvey was obliged to see patients at least once a week in the great hall, and to prescribe treatment for them.

A coin minted in Harvey's honor (right), and (below) St. Bartholomew's Hospital, London, in 1723. In that year a total of 4,163 patients entered the hospital, out of whom 3,381 were cured.

was made in the liver, and from there it traveled through the veins to every part of the body, nourishing the organs and tissues.

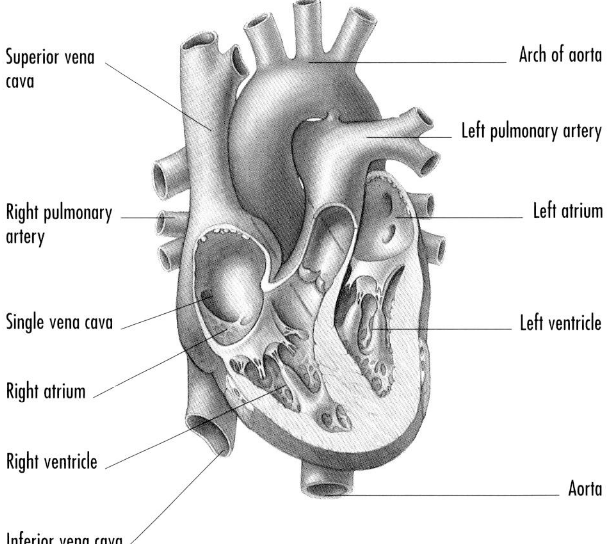

Harvey discovered that blood enters the right side of the heart through the superior and inferior venae cavae (which converge into a single vena cava), then passes through the pulmonary arteries to the lungs, where it receives oxygen. It returns to the left atrium, passes into the left ventricle, and is pumped out through the aorta to the rest of the body, eventually returning to the right atrium.

Galen's Theories of the Heart

Galen observed that blood entered the right side of the heart through the vena cava (Latin for "hollow vein"). He knew that somehow blood had to move from the right side of the heart to the left side, and decided that it must do so by passing through tiny pores, or openings, in the "septum," the narrow wall separating the two sides of the heart (see right). On the left side, he believed, blood mixed with air entering the heart from the lungs. This mixture left the heart through the aorta (the largest artery) and was carried by other arteries through the body. The heart and arteries possessed a "pulsive power" that made them expand, drawing in blood from the tissues next to them as they did so. Galen thought some blood was carried to the head and the brain. There the blood absorbed "animal spirit," which produced consciousness. The animal spirit was carried through the hollow nerves to the muscles and sense organs.

Galen, who served as personal physician to the Roman emperor Marcus Aurelius (121–180), was considered the foremost medical authority of his day. He obtained his ideas about anatomy through careful dissection (the cutting open of animals to learn more about their anatomy) and wrote more than 85 treatises on medical matters of all kinds. These remained the basis of all thinking about medicine and anatomy in Europe until the 16th century.

Matteo Realdo Colombo, professor of anatomy and surgery at Padua University, was one of the first to query Galen's teachings about blood. He did not think it traveled between the two sides of the heart through the walls of the septum as Galen said it did, and wondered if it flowed from the lungs to the heart, and perhaps also—by way of the pulmonary artery from the other side of the heart—back to the lungs. He saw that the heart alternately became bigger (dilated) and then smaller (contracted), and believed that it was the action of the heart as it contracted that pushed the blood into the lungs and out through the aorta.

Experimentation and Observation

Colombo's ideas were not widely accepted (he appears to have been extremely quarrelsome) but Harvey certainly knew of them when he began his own investigations into the heart and blood. Like Galen, Harvey believed in using dissection. He examined the hearts and blood vessels of nearly 130 mammals and identified the valves (Harvey called them "clacks") that separate the atria and ventricles (the chambers on the upper and lower sides of the heart; see above) and allow blood to flow in one direction only. He discovered that valves in the veins also allow blood to pass in only one direction—toward the heart.

The largest veins in the body are the superior and inferior venae cavae; these converge into a single vena cava as they enter the heart. Many of Harvey's experi-

The vivisection of a pig, as carried out by Galen and shown in a 1550 edition of his work. For many centuries the Greek physician was seen as an authority on anatomy, but Harvey disproved his theories on how the heart and blood worked.

Harvey's notebook with diagrams (above), and (left) his proof that blood returns to the heart through veins. The armband makes the veins stand out; the valves show as swellings. Because the valves within the veins only allow blood to flow one way, if pressure is put on a vein, blood will cease flowing toward the heart. A diagram (below) from Reisch's *Pearls of Wisdom* (1508) gives a fairly realistic picture of the main human internal organs: heart, lungs, liver, spleen, and bladder.

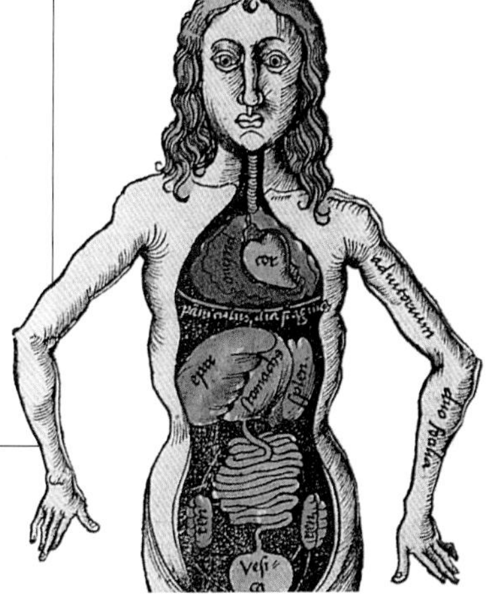

ments were carried out on live animals (this is called vivisection). When he tied the vena cava of a living snake he found that the vein bulged and the heart became smaller and paler in color. He concluded that blood must usually return to the heart through the vein, but he was blocking its path. When he tied the aorta, he found the heart swelled and became almost purple in color. The heart was swollen with blood; Harvey realized that the blood must usually leave the heart through the aorta, but he was preventing it doing so.

Harvey discovered that the ventricles of the heart are muscles that work as pumps, squeezing blood into the pulmonary artery, which carries it to the lungs, and into the aorta, which carries it to all parts of the body. As Colombo thought, it is the contraction (the "systole") of the heart that causes the arteries to pulsate, not the dilation (the "diastole") of the heart as had previously been supposed. You can feel your own pulse as a regular throb on the inside of your wrist, close to where you wear your watch. Each beat registers a contraction of the heart.

Harvey calculated that with every beat an adult's heart pumps about 4 cubic inches (60 cm^3) of blood.

This would mean the heart pumped about 68 gallons (259 liters) every hour, an amount of blood that would weigh more than 440 pounds (200 kg). It is clearly impossible for the body each hour to manufacture and consume a quantity of blood amounting to about three times the weight of the person. Harvey therefore concluded that a much smaller amount of blood circulates constantly through the body and is renewed and replenished as necessary.

In 1628 Harvey published his findings in a book called *Anatomical Exercises on the Motion of the Heart and Blood in Animals*. Galen was still very much admired and it was a brave man who dared to oppose his ideas. At first Harvey was scorned and ridiculed, and he was deserted by some of his patients. But by the time he died most doctors had come to accept his view of the circulation of the blood, and the influence of Galen soon declined.

Harvey accurately described almost every detail of the circulation of the blood. He observed that both arteries and veins divide repeatedly into ever smaller vessels, but he was unable to find the points where the arteries leading from the heart become veins, returning blood to the heart. Harvey did not have the use of a microscope. Four years after his death the Italian physiologist Marcello Malpighi (1628–1694), who did, was able to identify the network of delicate, thin-walled vessels called capillaries (from the Latin word for "hair") that connects the arteries and veins.

How a Chick Grows Inside the Egg

As well as his work on circulation, Harvey also conducted research into animal reproduction and the way embryos develop before birth (embryology). This was a subject about which little was known. Aristotle had suggested that the male seed produced the new animal; he thought the female's role was simply to nourish and protect the seed. Galen, on the other hand, believed that there were both male and female seeds, and that these united to produce the young animal. Harvey's investigations led him to conclude that all animals develop from a female egg, even those that are born as live young, including human beings. This was a bold statement that was not definitely proved until the German-Estonian biologist Karl Ernst von Baer (1792–1876), the founder of modern embryology, showed it to be true in 1827.

Harvey's friends encouraged him to publish his researches into embryos in his book *The Generation of Animals* in 1651. His studies were particularly concerned with the growth of the chick inside the hen's egg. From his observations, made with the naked eye, he was able to describe accurately the order in which the parts of the body appear in the developing chick: the heart is the first organ to become visible and the feathers appear last. He was also able to show that chick embryos go through the same stages of development as the embryos of mammals such as deer.

The *Generation of Animals* was less influential than Harvey's book on blood circulation. Soon after his death scientists such as Marcello Malpighi and Robert Hooke (1635–1703) were using microscopes to make much more accurate observations than Harvey had been able to do. Harvey had speculated that male semen might contribute to the formation of the embryo. However, spermatozoa (sperm cells) are not visible to the naked eye. The first person to see them was the Dutch microscopist Antoni von Leeuwenhoek (1632–1723), using the powerful microscope he developed.

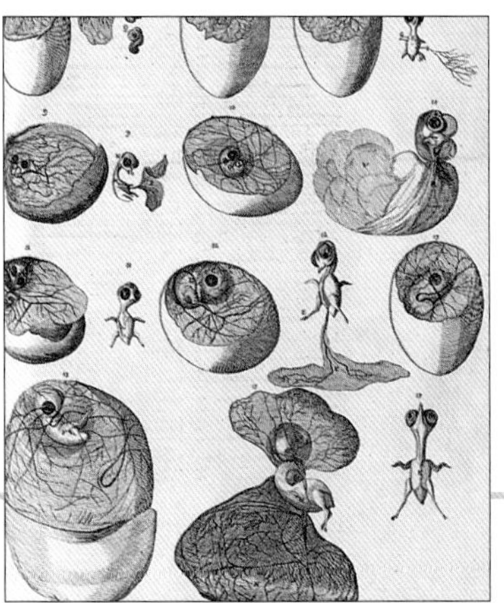

These detailed drawings from a work by Fabricius showing the development of a chick. He taught Harvey at Padua University, and prompted Harvey's own investigations into the development of animal embryos.

◀ **see also** Aristotle VOL 1:6
▶ **see also** Jenner VOL 2:16 Pasteur VOL 2:76 Fleming VOL 3:80 Salk VOL 4:78

HARVEY: Life and Times

SCIENTIFIC BACKGROUND

Before 1590

Greek physician Galen (129–c. 199 AD) discovers that arteries carry blood, not air

Greek physician Herophilus (active 300–250 BC) notes that arteries pulsate, but fails to connect this with the beating of the heart

Flemish anatomist Andreas Vesalius (1514–1564) publishes *On the Structure of the Human Body* (1543). It contains detailed descriptions of bones and the nervous system and disproves many of Galen's theories

1590

1600

1603 Hieronymus Fabricius (1537–1619), Italian anatomist and embryologist and teacher of Harvey, publishes *On the Valves in the Veins*

1609–1648 Harvey works as a physician at St. Bartholomew's Hospital in London

1610

1620

1620 *New Organon* by English philosopher Francis Bacon (1561–1626) stresses the importance of experimentation in scientific work

1621 Fabricius publishes work on embryology

1628 Harvey describes how blood circulates in his *Anatomical Exercises on the Motion of the Heart and Blood in Animals*

1630

1637 French philosopher Rene Descartes (1591–1650) publicizes Harvey's theories on the circulation of blood in his writings

1640

1650

1651 Harvey publishes his book on embryology, *The Generation of Animals*

1652 Danish physician Thomas Bartholin (1616–1680) describes the lymphatic system, defending Harvey's 1628 theory of blood circulation

1660

After 1660

1661 Using a microscope, Italian physiologist Marcello Malpighi (1628–1694) discovers blood capillaries

1665 English scientist Robert Hooke (1635–1703) publishes *Micrographia*, a collection of his observations using a microscope

1684 Dutch microscopist Anton van Leeuwenhoek (1632–1723) accurately describes red blood cells, bacteria and protozoa, and sperm cells

1827 German-Estonian embryologist Karl Ernst von Baer (1792–1876) describes the mammalian egg

POLITICAL AND CULTURAL BACKGROUND

1587 The Catholic Mary Queen of Scots (1542–1587) is executed on the orders of her cousin Queen Elizabeth I of England (1533–1603)

1593 The Irish rebel against English rule; the uprising lasts until 1601

1603–09 English playwright William Shakespeare (1564–1616) completes two of his greatest tragedies, *King Lear* and *Macbeth*

1608 French adventurer Samuel de Champlain (1567–1635) explores the North American coast and founds the city of Quebec in Canada

1619 The Dutch make Batavia (Indonesia) the center of their spice trade with the East Indies

1626 Governor Peter Minuit purchases the island of Manhattan from a local Native American tribe to found New Amsterdam (later New York)

1632 Dutch artist Rembrandt (1606–1669) completes the *Anatomy Lesson of Dr. Tulp*, which shows a body being dissected for medical research

1639 Japan's military governor Tokugawa Iemitsu (1604–1651) closes Japanese ports to foreigners in an effort to banish Christian missionaries

1648 The Thirty Years War, which has involved most countries in Europe, ends in the Peace of Westphalia

1649 The English Civil War ends with the beheading of King Charles I of Great Britain and Ireland (1600–1649) on January 30

1660 The monarchy is restored in Great Britain and Ireland as Charles I's son returns from Europe to be crowned King Charles II (1630–1685)

1665–66 The Great Plague strikes England, causing at least 70,000 deaths; the Great Fire of London that follows destroys many of the plague-carrying rats

ISAAC NEWTON

1642–1727

"Nature and Nature's Laws lay hid in Night: God said,* Let Newton be!** ***and All was Light."

ALEXANDER POPE (1688–1744)
Epitaph written for Newton

IT HAS BEEN WIDELY ACCEPTED SINCE ISAAC NEWTON'S OWN DAY THAT HE WAS THE GREATEST AND MOST IMPORTANT SCIENTIST OF ALL TIME. FOR THE PAST 300 YEARS HIS INSIGHTS, PARTICULARLY ON THE PHYSICS OF LIGHT, MOTION, AND GRAVITY, HAVE FORMED THE BASIS OF OUR UNDERSTANDING OF THE UNIVERSE AND OUR PLACE WITHIN IT.

ISAAC NEWTON'S BEGINNINGS GAVE LITTLE indication of how important he would become in the future. The son of a farmer in the village of Woolsthorpe in Lincolnshire, eastern England, Newton had a difficult early life. He was born prematurely on Christmas Day, 1642, about three months after his father's death, and was described as small enough to fit into a two-pint pot. When he was three years old his mother remarried and left Isaac with his grandmother while she moved to her new husband's home in a nearby village. She did not return until her second husband's death eight years later. Newton appears to have experienced some bitterness about this. One of his notebook entries confesses that he had threatened "my father and mother Smith to burne them and the house over them."

A FLAIR FOR INVENTION

As a youth, Newton revealed considerable skill in constructing working mechanical models. A windmill he built, for example, was operated by a mouse running on a treadmill. He also designed numerous sundials, developing in the process an ability to tell the time by the Sun with impressive accuracy. At one time he became so distracted by his design for a model waterwheel that he allowed the sheep he was supposed to be watching to stray into a neighbor's field.

As an only child, young Isaac inherited his father's farm; it was assumed that he would take over the running of it, as his mother wished him to do. However, the boy clearly had other interests and, in 1661, at the suggestion of her brother, Newton's mother agreed instead to send her son to Trinity College, Cambridge. Here Newton developed a particular passion for mathematics and astronomy. As at other universities, the

It was supposedly while sitting in his garden at Woolsthorpe that Newton began to wonder whether the same force that makes apples fall from trees keeps the Moon in orbit around the Earth. The scene is recreated in this 19th-century painting.

Woolsthorpe Manor, Lincolnshire, Newton's family home. The house still stands today and is preserved as it would have been in Newton's time.

KEY DATES	
1642	Born December 25 at Woolsthorpe Manor, Lincolnshire
1645	Remarriage of mother Hannah to Reverend Barnabas Smith
1653	Mother returns to Woolsthorpe after death of second husband
1655	Attends King Edward VI Grammar School, Cambridge
1661	Admitted into Trinity College, Cambridge
1672	Publishes *A New Theory on Light and Colors*
1679	Death of mother
1687	*The Mathematical Principles of Natural Philosophy*, known as the *Principia*, describes Newton's theory of planetary motion
1693	Suffers nervous breakdown
1696	Appointed warden of Royal Mint in London; becomes master of the Mint in 1699
1703	Elected president of the Royal Society, London
1704	Major work on light and color, *Opticks*, published
1705	Knighted by Queen Anne to become Sir Isaac Newton
1727	Dies March 20; later buried in Westminster Abbey

curriculum was based principally on Aristotle's ideas, but Newton began to study independently the works of more modern philosophers such as the Englishman Francis Bacon (1561–1626) and the Frenchman René Descartes (1596–1650).

PRODUCTIVE YEARS

By the summer of 1665 the Great Plague, which had broken out in London and eventually killed more than 75,000 people, had reached Cambridge. The university was closed. Newton returned home to Woolsthorpe and spent the whole of the next year there. The year 1666 is sometimes described as Newton's year of wonder ("annus mirabilis"). He was, he said later, at the height of his inventive powers, and more in love with mathematics and philosophy than at any other time of his life. He first developed his theories of color and gravity at this time, as well as a range of new mathematical ideas. Yet he kept much of this work to himself, showing just a part of it to colleagues later. The rest was only published after his death.

By 1669 Newton had returned to Cambridge and become professor of mathematics there. His abilities began to be recognized when he designed a radical new telescope, which he presented to the Royal Society, later to become the world's leading scientific institution. This so impressed the Society that in 1672 they elected him a member.

EXPERIMENTS IN LIGHT

In the same year Newton wrote a letter to the Royal Society setting out *A New Theory on Light and Colors.* It met with considerable criticism, particularly from the distinguished physicist, Robert Hooke (1635–1703). Newton was deeply affected by this and returned

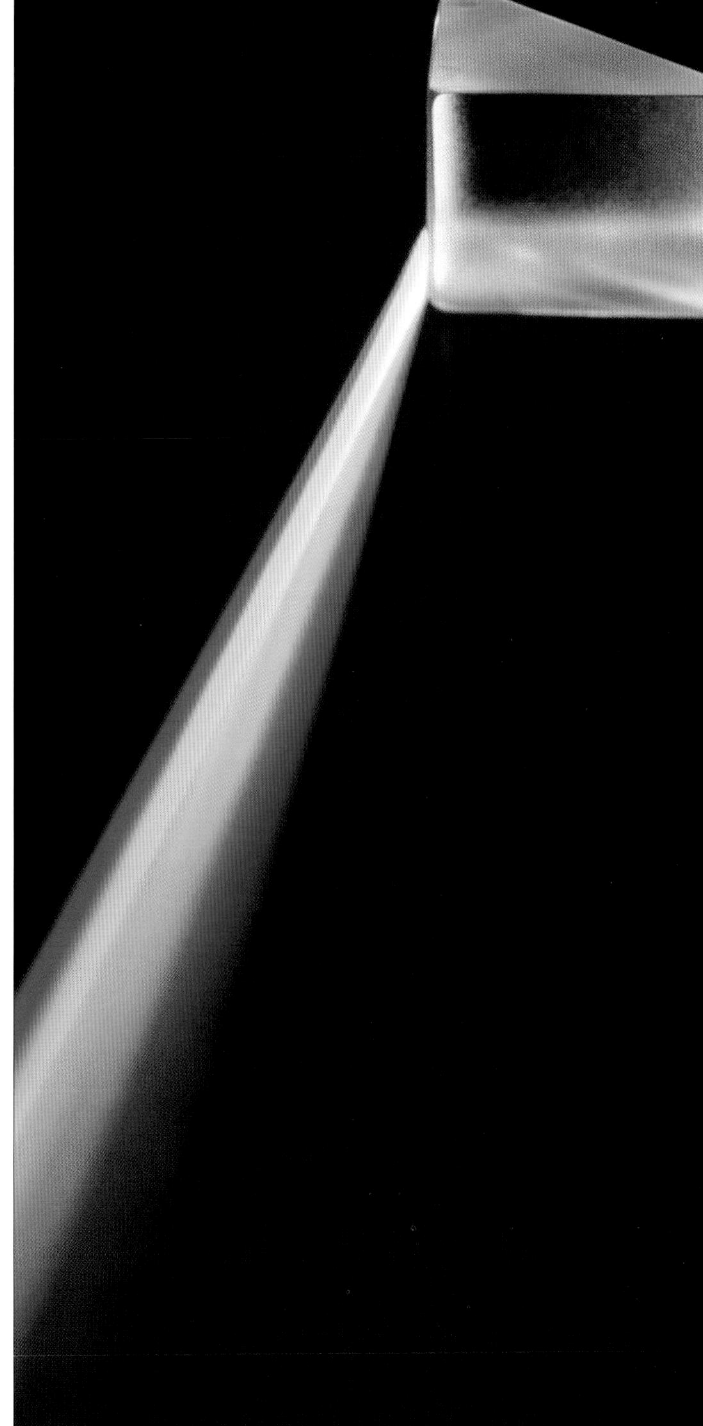

This picture confirms what Newton's experiments revealed. As white light passes through a prism—a triangular block of pure glass—the beam divides into the colors of the spectrum. The blue end of the spectrum is refracted most strongly, and the red least; we now know that this happens because each color band has a unique range of wavelengths.

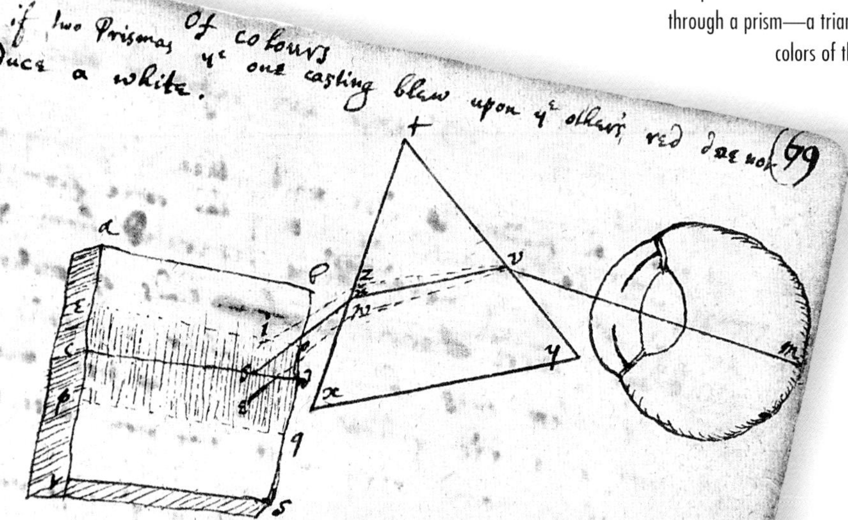

A sketch by Newton illustrates how light is refracted by a prism. His work on colors and light was fully described in his book, *Opticks* (1704).

to working on his own; after this he was always slow to reveal his discoveries and reluctant to publish his findings.

Newton continued his experimental work with color, however. At this time, scientists did not believe that color is a basic property of light, but thought it was a modification that occurred when light passed through another substance such as water, cloud, or glass. Newton, however, became intrigued by his observations of what happened when light passes through prisms—triangular blocks of glass.

When light rays pass between substances of different densities, such as air and water, they change direction. This is called refraction. It is refraction that makes a stick half in and half out of water look as if it is bent or broken. Newton noted that a thread, colored half red and half blue, when pulled tight and viewed through a prism, seemed to split in half; the blue half appearing a little higher than the red (see below right). Newton wondered whether the rays that made blue appeared higher than those that made red because they were refracted more.

THE COLORS OF THE RAINBOW

Newton tested his idea with further experiments, and in 1672 made a series of groundbreaking observations. In a darkened room he passed a beam of light through a prism onto a screen. As the light traveled through the prism it fanned out into the colors of the spectrum. These are the colors seen in a rainbow, ranging from red at one end to violet at the other. Newton demonstrated that if a beam from a single color from the spectrum—red, for example—was passed through a second prism, this light remained red: it could not be changed again. When the second prism was replaced with a lens and the colors of the spectrum passed through this, they turned to white light again (see below).

Newton drew a number of conclusions from these experiments. He decided that color must be a basic component, or part, of light, and that white light was the result of combining all the colors of the spectrum. He also found that every color in the spectrum has a different and unique degree of refraction. We now know that this is because light is made up of waves. Each color band has a unique range of wavelengths, and each color band is always refracted the same amount by a prism. Red light, for instance, has a longer wavelength than violet light, and so is refracted less. Newton, however, thought that light was made up of small particles that he called "corpuscles." These, he suggested, were ejected from the source of light like bullets from a gun.

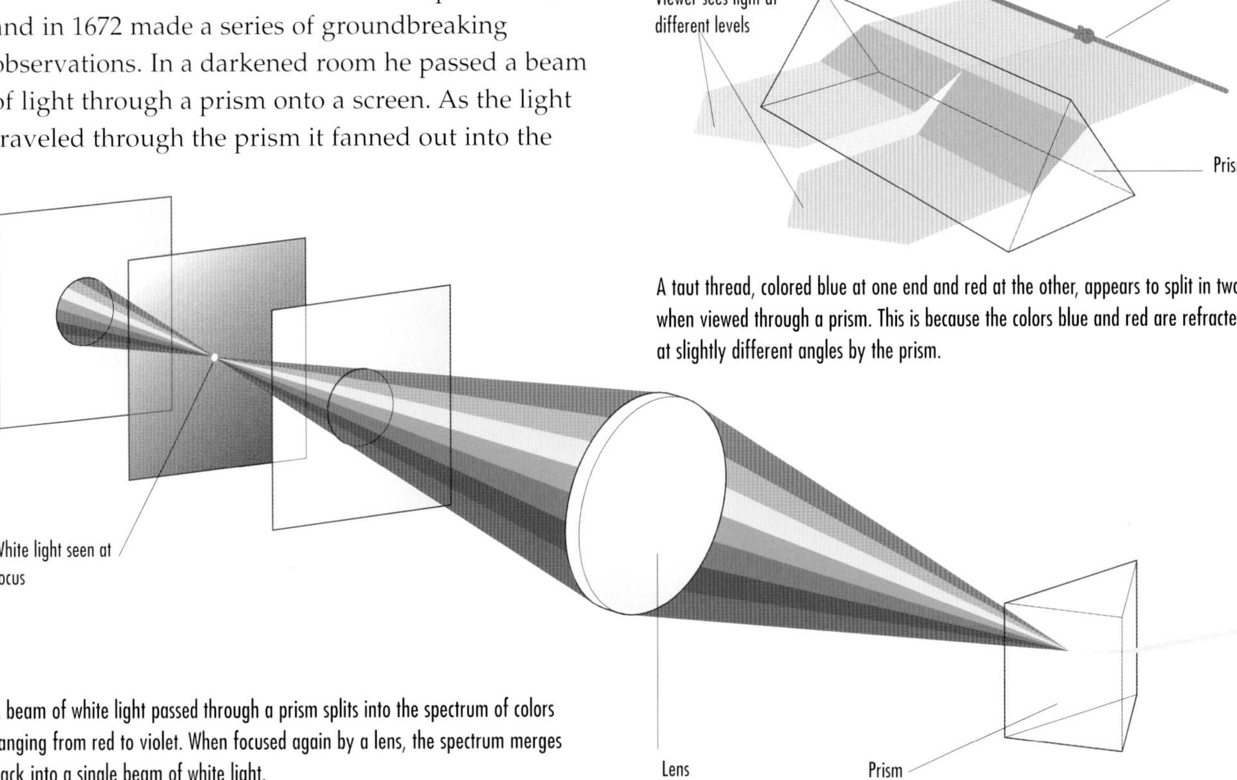

A taut thread, colored blue at one end and red at the other, appears to split in two when viewed through a prism. This is because the colors blue and red are refracted at slightly different angles by the prism.

A beam of white light passed through a prism splits into the spectrum of colors ranging from red to violet. When focused again by a lens, the spectrum merges back into a single beam of white light.

Newton's views on the nature of light were challenged by Robert Hooke, who had also made a study of optics, and by the Dutch physicist Christiaan Huygens (1629–1693). They both favored the wave theory of light, but Newton firmly resisted their arguments. He pointed out that while we know that sound waves are able to bend because we can hear a bell ringing from the other side of a hill, light does not behave in the same way—the hill prevents us from seeing the church tower in which the bell is ringing. In fact, we now know that light sometimes behaves like particles and sometimes like waves.

As so often in his life, Newton was angered by criticism of his views. His reaction to Hooke's and Huygens' objections was extreme: he refused to publish anything more. Not until 1704, more than 30 years later, did the results of his work on light and color appear—with some additions and revisions—in his book *Opticks*. Meanwhile, he had turned to consideration of how things move.

PERPETUAL MOTION

Questions about motion had kept thinkers busy for centuries, but no one had yet come up with a satisfactory explanation to replace Aristotle's theory that things move because they are pushed or pulled by something or someone, a "mover," (see page 10). French philosopher Jean Buridan (c. 1300–1385) had proposed the idea of impetus, which he conceived as a driving force imparted to the object by the mover, to explain why objects continued to move through the air until slowed down and brought to a halt by friction or air resistance.

> *"Everything doth naturally persevere in that state in which it is, unless it be interrupted by some external cause."*
>
> FROM NEWTON'S STUDENT NOTEBOOK OF 1664

Newton, strongly influenced by the Italian mathematician Galileo Galilei (1564–1642), and by René Descartes, rejected both these theories. As far back as 1664 he had commented in a student notebook: "Everything doth naturally persevere in that state in which it is, unless it be interrupted by some external cause [everything remains naturally in the state in which it is, unless interrupted by an outside force]." This tendency is known as "inertia." An object at rest will remain at rest until something or someone moves it. No one expects a book lying on a level table to move or jump into the air of its own accord. Newton applied the same reasoning to moving objects. An object moving in a straight line cannot stop, change direction, speed up, or slow down of its own accord. Just as an outside force is needed to begin motion, so an outside force is required to halt it. Without this force, Newton believed, the object would continue in "its present state" endlessly. This principle would later be formulated as his first law of motion.

What are the outside forces that slow objects down and bring them to a stop? Clearly an object will cease moving if a solid obstacle is placed in its path, while it had long been recognized that friction and air resistance will end the movement of objects propelled through the air, such as arrows. Newton, however, went on to argue that, if you could remove friction and air resistance, an object that has been set in motion will continue moving. It is inertia that will keep the object in motion until a greater force acts upon it. Had he been able to observe movement in space, he would have seen

Newton in Space

The launching of a space rocket involves all Newton's laws of motion. The engines must overcome the inertia of the rocket (as stated in his first law) and the pull of gravity. As described in his third law, the burning fuel in the engines produces equal and opposite reactions, the force of the expelled gases providing thrust to lift the rocket (right). This law is also demonstrated in an illustration from an 18th-century textbook on Newton (below). Newton's second law notes that the less the mass of the object, the greater the acceleration. The rocket loses mass as it burns fuel and so accelerates accumulatively.

The principle that to every action there is an equal and opposite reaction is also shown by the toy known as Newton's cradle (left). If one ball is swung against the others, only one ball will rise on the other side; if two balls are set in motion, two of the balls opposite will rise, and so on.

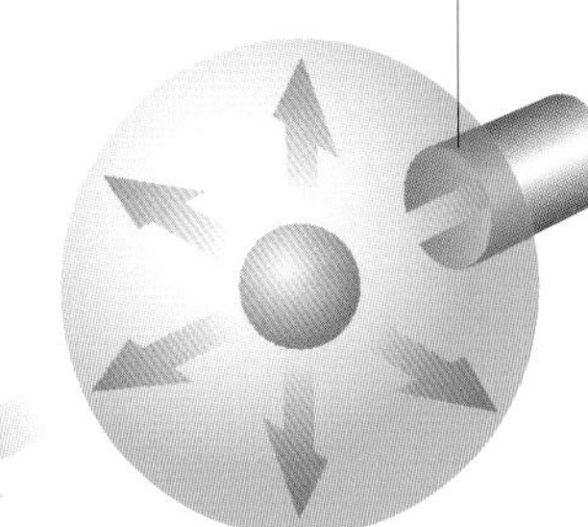

his theory proved. Because space contains no atmosphere (air), there is nothing to resist movement, and a single push on an astronaut who is not attached to a spaceship will condemn him or her to move through space for ever.

The French mathematician René Descartes (1596–1650), one of the greatest figures in the history of Western thought and undoubtedly the most influential thinker of his day, had already touched on many of these ideas in his writings, which were well known to Newton. But it was Newton who realized that it would be possible to formulate precise and exact laws to apply to all forms of motion, whether on Earth or in space.

Newton's Three Laws of Motion

Newton eventually succeeded in formulating three laws of motion, which he wrote up in 1687. As already noted, the first states that "a body continues to move with a constant velocity [speed and direction] or to…remain at rest unless acted upon by an external force." In other words, a moving object continues to move at a steady speed in a straight line, or if at rest continues to stay at rest, unless interrupted.

The second law states that when a force is applied to an object it accelerates in a straight line in the direction of the force. The object's acceleration (the rate of change in its velocity) depends on its mass (the amount of matter it contains). The less the object's mass, the greater the acceleration (a light stone is easier to throw than a heavy one).

Newton's third law maintains that if an object exerts force on another object (the action), then that second object exerts an equal but opposite force back (the reaction). If you push or pull an object it will push or pull you back to an equal extent. So when an astronaut pushes against a space ship with a certain force, the space ship pushes against the astronaut with an exactly equal force.

In the 20th century it was discovered that Newton's laws of motion do not apply in the area of quantum mechanics (the physics of very small-scale or high-speed particles and atoms). But the laws of motion formulated by Newton more than 300 years ago hold good for all objects moving at normal speeds.

KEPLER'S PUZZLE

In 1609 Johannes Kepler (1571–1630) had shown that the planets move in elliptical orbits (see page 42). In such an orbit a planet finds itself at varying distances from the Sun and moves with varying speeds—fastest when it is at its closest point (perihelion) to the Sun, and slowest when farthest away (aphelion). By Newton's time many scientists had attempted to account for this behavior, but no satisfactory explanation had yet been found. It was generally assumed, however, that such changes in speed were due in some way to the Sun.

After the response to his theories on light and color, Newton had been unwilling to publish much of his work, and it was known only to a handful of mathematicians. However, Newton's life changed crucially after he was visited by the astronomer Edmond Halley (1656–1742) in August 1684. The visit was prompted by a discussion with fellow scientists Robert Hooke and Christopher Wren (1632–1723) in which all three had failed to solve the Keplerian puzzle of what held the planets in orbit. Halley came to see if Newton could produce a solution. Newton claimed to have found the answer but needed time to check his mathematical proof. This led him to consider additional queries and problems, and eventually his original short proof had expanded into a volume, published three years later, in 1687, as *The Mathematical Principles of Natural Philosophy*. This became known as the *Principia* from its Latin title. The book described the workings of the universe with great mathematical precision, and had an extraordinary impact on the scientific world.

Newton's Telescope

Telescopes were a vital tool in the quest for scientific knowledge. However, early telescopes used spherical convex (rounded) lenses to bring the light from the object being observed into focus, and this caused a problem. Because single spherical lenses do not refract light of different wavelengths equally, these lenses made the observed objects look blurred and gave them a colored fringe; this is known as chromatic aberration.

Newton's solution was to replace the lens with a concave (hollowed) mirror. Light entering the telescope was reflected onto another, smaller mirror angled at 45° to the tube's axis. This directed it into an eyepiece on the side, which magnified the image. In 1672 French scientist N. Cassegrain invented a telescope in which a secondary mirror reflected light back through a hole in the main mirror. Both types of telescope are still used today.

Newton's reflector telescope (above and below) used mirrors to reflect the image into the eyepiece on the side.

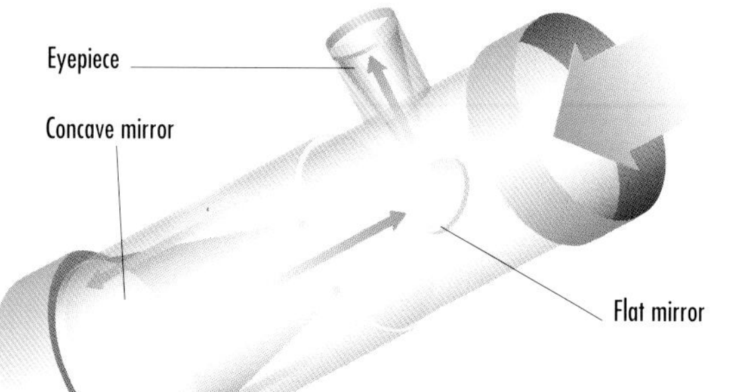

GRAVITATIONAL PULL

In outline, Newton's assertions were incredibly simple. First he stated that there was only one force at work—a force of attraction. This is what we call the force of gravitation, and Newton showed that it governs all motion, whether in space or on Earth. Every piece of matter in the universe attracts every other piece. The force of gravitation operates in a surprisingly neat way, depending only upon the masses of the bodies involved and the distance between them. All other factors can be ignored. It does not matter if one object is hot, the other cold, or if one is made of iron, the other of copper. As a further simplification Newton demonstrated that the attractive force of a spherical object was concentrated at the center of the sphere.

Edmond Halley
1656–1742

Edmond Halley was an English astronomer and mathematician who has a remarkable catalog of achievements to his name but is best remembered for his contribution to the astronomy of comets; he first observed the comet that bears his name in 1682. Halley was also the founder of modern geophysics, making discoveries about the Earth's magnetic field, atmospheric pressure, and the aurora borealis (the bands or streamers of lights seen in the northern sky under certain conditions). He published the first catalog of southern stars to be located by telescope in 1678, and the first meteorological chart in 1686. Halley persuaded Newton to publish his major work on mathematics, *Principia*, and organized its distribution and sale.

In later life Newton reported that the basic theory had come to him while he was sitting in the orchard of his home at Woolsthorpe in 1666, his year of wonders. The famous story is that he watched an apple fall to the ground. He suddenly realized that the power of gravity that brought the apple down need not be limited to any particular distance from the Earth but might extend much farther. "Why not as high as the Moon?", he is supposed to have asked. Newton realized that just as the Earth's gravitational pull attracts an apple, making it fall to the ground, so it attracts the Moon and keeps it in orbit.

How does gravity cause the Moon to orbit the Earth? As Newton had shown by his first law of motion, if the Earth were not there then the Moon would move through space in a straight line and at a steady speed. However, a balance is achieved between the Moon's being pulled down toward the Earth by Earth's gravitational pull and the Moon's forward motion along its original path. It is this balancing process that deflects the Moon sideways and keeps it in orbit around the Earth.

As with the Earth and Moon, so with the Sun and the planets. Newton argued that the force between any two bodies (the Sun and a planet) varies depending on their mass and the distance between them. He calculated that the attractive force of gravity increases with mass and weakens with distance, operating according to an "inverse square" law. So a planet twice as far away as another would be pulled with a quarter of the

The room at Woolsthorpe Manor in which Newton was born. A copy of his most famous work, *Principia*, lies open on the desk, and a reflector telescope similar to the one that Newton constructed stands on the windowsill.

force, a planet four times as far away with one-sixteenth of the force. This inverse square law allowed Newton to prove mathematically what Kepler had observed, that the planets move in elliptical orbits. The French writer and Enlightenment figure Voltaire (1694–1778) was a great admirer of Newton and helped to publicize his theories in France. He wrote: "Before Kepler all men were blind. Kepler had one eye, and Newton had two eyes," meaning that Kepler saw half the truth about the Universe, but Newton saw it whole.

Tracking Comets

Further support for Newton's theory was provided by comets. A comet is a small celestial object consisting of a nucleus (made up of ice and dust particles) and a long luminous tail pointing away from the Sun. From ancient times comets had seemed to appear suddenly and unpredictably, flashing quickly and brightly across the heavens, never to be seen again. However, Newton and Halley began to suspect that comets, like the planets, move around the Sun in elliptical orbits. Because they travel in larger orbits than the planets, they are observed less frequently from Earth, but Newton and Halley were sure that their orbits would prove to be as predictable as those of the planets.

In 1682 Halley tracked the path of a comet as it moved across the sky and checked it against the astronomical records of comets observed in 1531 and 1607. All three paths were very similar, and Halley decided that each occurrence must be the same comet returning close to Earth on its orbit around the Sun. He predicted, correctly, that the comet would return again in 1758–59. When it did appear, 17 years after his death, it was

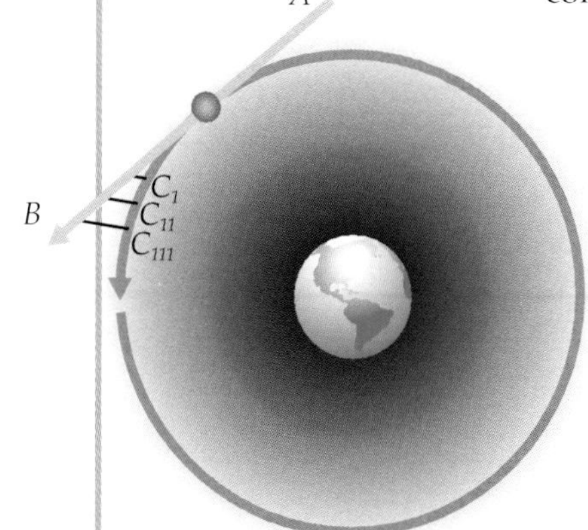

According to Newton's law of inertia, the Moon would naturally move in a straight line, from A to B. Because it is attracted by the Earth through its gravitational pull, the Moon instead moves to points C′, C″, and C‴. This establishes its orbit around the Earth.

A Society for the Study of Science

The Royal Society, still the foremost institution for the encouragement of scientific understanding in Britain, was founded during Newton's lifetime when, on November 28, 1660, a group of learned men met in London to hear a lecture given by Christopher Wren (1632–1723). Although he is better known today as an architect, Wren was a mathematician and a professor of astronomy. Afterwards, those attending the lecture decided to found "a College for the Promoting of Physic-Mathematicall Experimentall Learning." Members (called "fellows") assembled each week to watch experiments being carried out and to discuss scientific issues. When King Charles II (1630–1685) heard about the group, he granted it a royal charter in 1662 and gave it the name of the Royal Society of London for Improving Natural Knowledge. Today it is known simply as the Royal Society, and still exists to encourage the understanding of science, the promotion of scientific experimentation, and the discussion of results throughout Britain and the world.

Experiments in Science

Although founded sometime after his death, the Royal Society was the brainchild of English philosopher and politician Francis Bacon (1561–1626). It was he who first suggested setting up an academy for scientists in England in his book *The New Atlantis*, published the year after his death. Bacon was a keen exponent of scientific experimentation. He jumped from his carriage one winter's night in order to pack a chicken's carcass with snow to see if the extreme cold would preserve the bird's flesh—an early, but fatal, experiment in deep-freezing; Bacon subsequently died from the effects of the cold. The first "curator of experiments" at the Royal Society was Robert

The Octagon Room in the Royal Observatory at Greenwich (above). The Greenwich Observatory was founded in 1675 and presided over by John Flamsteed, the first astronomer-royal of England, and another of the scientists with whom Newton had a dispute.

The frontispiece of Thomas Sprat's 1667 History of the Royal Society *(left) glorifies the scientific advances of the age. King Charles II gave royal approval to the Society; his statue is shown being honored with a laurel crown. On the left of the column is the philosopher Francis Bacon; on the right is William, Viscount Brouncker, who was—from 1662 to 1677—the Society's first president.*

Hooke. A versatile scientist, Hooke developed a compound microscope that allowed him to study tiny objects invisible to the naked eye. His *Micrographia* (1665), containing his findings, was one of the first works published by the Royal Society.

Newton and the Royal Society

Newton became a member of the Royal Society in 1672. When, in 1686, he delivered the manuscript of the first volume of his *Principia* to the Society, Hooke used the occasion to renew his attack on Newton, pointing out that he had himself already anticipated many of Newton's ideas on motion and gravitation. Newton never forgave Hooke and refused to accept the presidency of the Royal Society until after Hooke's death in 1703. Although Newton worked actively to raise the public status of science, he also used the post of president (which he held for 24 years) to air his disagreements with scientists such as John Flamsteed (1646–1719), the astronomer-royal who had compiled the accurate observations of the Moon needed by Newton for his work. He was reluctant to part with them, and Newton exploited his influence with the royal court to have him give them up.

named "Halley's Comet" in his honor. Halley's Comet takes 76 years to orbit the Sun. Halley's Comet last passed close to the Earth, traveling at a speed of more than 80,000 mph (128,000 kph) in 1986. The last bright comet visible from Earth with the naked eye was the Hale-Bopp comet in 1997, first identified by amateur astronomers Alan Hale and Thomas Bopp. It had last reached its perihelion (its closest point to the Sun) some 4,200 years ago.

Pages from Newton's notebook show how he solved a series of complicated mathematical problems using the calculus.

FINDING THE ETHER

Many people had difficulty in accepting certain aspects of Newton's science. For example, they could not understand how a force such as gravity could act over the vast distances of space. Newton recognized that he needed to find an answer to this question if his theory of universal gravity was to gain widespread acceptance, but he himself could not explain it satisfactorily. At one stage he simply declared that the search for the causes and mode of operation of gravity was unnecessary. "It was enough," he said, "that gravity really exists…." He insisted in a famous phrase that he would "frame no hypotheses [put forward no suggested explanations]" because "hypotheses… have no place in experimental philosophy."

Robert Hooke
1635–1703

Robert Hooke was a highly gifted English scientist and inventor, whose interests stretched from astronomy to microscopy. In 1660, after studying the elasticity of solids, he formulated Hooke's Law, which defined the mathematical relationship between stress and strain. Hooke applied his findings to a design for spring balances in watches. In 1662 he became the first curator of experiments at the Royal Society, and in 1665 published *Micrographia*, which featured microscopic studies and illustrations of structures such as snowflakes. His belief that the attractive force of gravity weakens according to an "inverse square" law led him to claim he had anticipated Newton's laws of planetary motion, and to an acrimonious dispute that lasted until Hooke's death in 1703.

Another time, he said that it was absurd to suppose that gravity could act at a distance "through a vacuum without the mediation of anything else [without the intervention of any other medium]." In other words, he was coming to believe in the existence of the ether, or quintessence, filling all of space. This was the transparent universal medium believed in by ancient and medieval philosophers. Aristotle had supposed it to be a rigid transparent substance; later thinkers such as Descartes thought it was a fluid that could transmit forces as readily as waves were transmitted through water. Helpful though such a medium might be to Newton's immediate problem, it also raised a fresh set of problems. If, for example, the ether was sturdy enough to transmit gravitational forces, it would surely offer strong resistance to the free movement of the planets, causing them to slow down.

In 1717 Newton devised an experiment to test the existence of the ether. When he placed a thermometer in each of two sealed glass vessels, one of which contained a vacuum, he found that the thermometer in the vacuum warmed up as much, and almost as quickly, as the other one. Newton concluded that this meant that heat was transmitted through the ether, which was more rare and subtle than air. Belief in the ether did not disappear until the 19th century when physicists Albert Michelson (1852–1931) and Edward Morley (1838–1923) proved it did not exist in the course of experiments to find how the motion of the Earth affects light waves.

NEWTON AND MATHEMATICS

All Newton's discoveries were based on his superb grasp of mathematics. His collected mathematical papers run to eight large volumes and cover all the main areas of the subject. An early indication of his

Newton and the Enlightenment

The Enlightenment, or the Age of Reason, is the name historians give to the period in the late 17th and 18th centuries when the study of science and philosophy was seen to offer a more rational basis for understanding the world than traditional religion. Inspiration for these ideas came from the writings of French philosopher René Descartes (1596–1650), who conceived the possibility of a unified system of knowledge and truth based on mathematics and reason. Two Englishmen were especially influential in the spread of Enlightenment thought in the late 17th century: philosopher John Locke (1632–1704), whose *Essay Concerning Human Understanding* (1690) argued that we only obtain knowledge of the world around us via our senses, and Isaac Newton.

In his *Principia* (1687) Newton succeeded in showing that the universe had a logical structure that could be properly understood by the application of mathematics. Although Newton's published writings were mainly concerned with exploring specific scientific problems, they opened up a new way of looking at the world. His ideas were popularized in France by Voltaire (François Marie de Arouet, 1694–1778), the greatest figure of the French Enlightenment, where they contributed to the intellectual and political discussions of those questioning the authority of Church and state in the years before the French Revolution of 1789.

Fundamental to this debate was the need for a system of checks and balances in government (as opposed to the absolute power exercised by kings). Locke, the principal author of this concept, derived his arguments from Newtonian philosophy. They influenced Benjamin Franklin (1706–1790) and the other authors of the American Declaration of Independence (1776) and were eventually incorporated into the U.S. Constitution.

In a famous portrait, the English poet and visionary artist William Blake (1757–1827) shows Newton gazing down to Earth as he measures the heavens. Blake disliked Newton's mechanistic view of the universe.

mathematical interests are found in a manuscript of 1665 (when he was living at Woolsthorpe), in which he calculated the area under a hyperbola to 55 decimal places, a painstaking exercise that had no practical or theoretical value. Another significant finding was in the field of algebra. Mathematicians before Newton were unable to deal with a particular sort of algebraic problem where the answer was not a whole number. In 1676 Newton published a formula that allowed all such values to be worked out.

Newton's major contribution to mathematics, however, was his development of an early form of the calculus, the branch of mathematics dealing with continually changing quantities. A similar system was developed at much the same time by the German mathematician Gottfried Liebniz (1646–1716). This gave rise to an acrimonious dispute with Newton about who had arrived at it first. It is now thought that Newton discovered the technique in the 1660s, while Leibniz arrived independently at the same conclusions in the 1670s. However, Leibniz was the first to publish his results. Jealous as ever for his reputation, Newton insisted that Leibniz had somehow stolen his work, and the argument was not resolved during Newton's lifetime. Leibniz's more precise notation was adopted in Europe, with the result that use of the calculus developed much more fully there than in Britain during the 18th and 19th centuries.

Kepler had earlier used a method similar to the calculus in trying to work out what volume of wine a curved barrel held (see page 43). He imagined the barrel to be made up of a great number of very thin

Gottfried Wilhelm Leibniz
1646–1716

German philosopher and mathematician Gottfried Leibniz was a highly original thinker who made important contributions across an extraordinary range of subjects: science, history, law, philosophy, politics, and theology. He invented a calculating machine; he also constructed a vision of a cosmos made up of simple, indestructible elements called monads. His theory of infinitesimal calculus was published in 1684, prompting his long-running dispute with Newton. His essays on theology placed faith in enlightenment and reason, a view satirized by the French writer Voltaire (1694–1778), who represented this view in *Candide* as the belief that "all is for the best in the best of all possible worlds."

circles of wood. By calculating the area of each circle, Kepler was able to add them together to arrive at a total volume. The use of these minutely small, or infinitesimal changes was central to the method behind the calculus, which was used to solve a range of mathematical problems. The techniques developed by Newton allowed him to make complicated calculations of curves and areas.

Newton's mathematical skills were put to the test in 1696 when the Swiss mathematician Jean (or Johann) Bernoulli (1667–1748) delivered a challenge to "the most acute mathematicians of the world" to solve a particular problem within six months. Newton is known to have received the problem at 4 o'clock in the afternoon. He had solved it by 4 o'clock the following morning. Even though Newton published his solution anonymously, Bernoulli had no doubts at all about who the author was. "The lion," he declared, "can be recognized by its footprint."

A 19th-century drawing shows Newton (seated at left) presiding at a meeting of the Royal Society. In later years, Newton was much occupied with public affairs, but he also found time to write on many aspects of science, and on theology.

ILLNESS AND PUBLIC SUCCESS

In 1693 Newton appears to have suffered a serious breakdown. Some historians think his symptoms may be an indication of mercury poisoning. Mercury, a highly toxic metallic element that can cause dementia, was much used in alchemy, the precursor of modern chemistry. Newton—whose interests embraced all branches of science and philosophy—is known to have studied alchemy in great depth and to have conducted chemical experiments throughout his life. However, others point to stress in his private life as a more likely cause of his breakdown. He appeared to observers to be "in a frenzy," accusing his friends of plotting against him and trying to entangle him with women. The illness, whatever its cause, seems to have passed within the year.

A coin depicting a stern-faced Newton commemorates him in classical style. Although his post at the Mint was essentially honorary, Newton chose to take a keen interest in driving through recoinage and taking action against counterfeiters.

By now it was clear that Newton desired to see his career marked by appointment to a significant public position. In 1696 an influential friend was able to secure for Newton the post of warden at the Royal Mint, the body responsible for producing all the English coinage and medals. He remained at the Mint for the rest of his life, being promoted to master in 1699.

These were busy years at the Mint, as England was undergoing a great reform of its coinage. The old silver coins had a smooth edge, making it easy for people to clip small amounts of silver from them and still be left with usable coins. Counterfeiting (forging false coins) was also common. To avoid both these problems, it was decided to put new coins with milled or grooved edges into circulation. Newton oversaw the change from the old to the new coinage; he also took a leading role in the campaign against counterfeiters, visiting taverns and prison cells to take statements from prisoners and informers. Newton carried out his duties with ruthless efficiency: in 1697 alone, 19 counterfeiters were executed under his orders.

Newton served two terms as an English member of Parliament, and was elected president of the Royal Society in 1703. He was knighted in 1705 by Queen Anne for his public duties. In the last years of his life he published further editions of his principal works, with some additions and alterations, but he did not undertake any major new scientific work.

"If I have seen further [than others]," Newton once said, "it is by standing on the shoulders of giants." On another occasion, he summed up his achievements in a striking image: "I do not know what I may appear to the world, but to myself I seem to have been only like a boy playing on the seashore, and diverting myself in now and then finding a smoother pebble or a prettier shell than ordinary, whilst the great ocean of truth lay all undiscovered before me…."

◀ **see also** Galileo VOL 1:24 Kepler VOL 1:38
▶ **see also** Einstein VOL 3:56 Hubble VOL 4:14 Hawking VOL 5:26

NEWTON: Life and Times

SCIENTIFIC BACKGROUND

Before 1660

Greek philosopher Aristotle (384–322 BC) argues that things on Earth move when pushed or pulled by a "mover"

German astronomer Johannes Kepler (1571–1630) publishes his first two laws of planetary motion in *The New Astronomy* (1609), revealing that planets move in elliptical paths

Italian mathematician, physicist, and astronomer Galileo Galilei (1564–1642) publishes his *Discourses upon Two New Sciences* (1638), in which he presents his laws of motion and friction, contradicting many of Aristotle's assertions

French philosopher René Descartes (1596–1650) sets out his view of the cosmos in *Principles of Philosophy* (1644)

1660

1662 The Royal Society in London is granted its royal charter

1665–66 Newton develops ideas on calculus, light and optics, and gravity

1668 Newton invents the reflecting telescope

1670

1672 English physicist Robert Hooke (1635–1703) suggests a "wave" theory of light

1675 English astronomer John Flamsteed (1646–1719) is appointed astronomer-royal at the Greenwich Observatory in London, England by King Charles II (1630–1685)

1679 Hooke suggests that gravitational force might operate according to an "inverse-square" law of attraction, leading to a later dispute with Newton

1680

1684 German mathematician Gottfried von Leibniz (1646–1716) invents calculus, a branch of mathematics, prompting a long-running argument with Newton

1687 Newton presents his three laws of planetary motion and his law of universal gravitation in his groundbreaking book, the *Principia*

POLITICAL AND CULTURAL BACKGROUND

1665 The New Jersey colony is founded by English colonists, who make Elizabethtown their capital

1675 Work begins on St. Paul's Cathedral in London, England. It is designed by English architect, astronomer, and mathematician Christopher Wren (1632–1723)

1682 English religious nonconformist William Penn (1644–1718), member of the Society of Friends (Quakers), is granted land in North America by King Charles II to establish a Quaker colony; the territory is called Pennsylvania

1683 William Penn signs the Great Treaty of Shackamaxon, by which the Delaware Native Americans grant him vast territories

1690

1690 Dutch physicist Christiaan Huygens (1629–1693) publishes his theory of gravity in *Discourse on the Cause of Gravity*

1690 Huygens publishes his *Treatise on Light*, which has been almost complete since 1678. In it he explains reflection and refraction, and puts forward his wave or pulse theory of light

1690 William III of England (1650–1702) completes the Protestant conquest of Ireland when he defeats the Catholic ex-king of England, James II (1633–1701), at the Battle of the Boyne

1690 The English philosopher John Locke (1632–1704) produces his *Two Treatises of Civil Government,* which supports natural law above the authority of a ruling body

1697 Tsar Peter the Great of Russia (1672–1725) visits Europe unannounced; the trip encourages him to Westernize his own country

1700

1704 Newton rejects the wave theory of light, presenting his work on light and color in his book *Opticks*

1705 English astronomer Edmund Halley (1656–1742) applies Newton's ideas to comets, correctly predicting the return of one—later to be called "Halley's comet"—on its orbit around the Sun

1703 Tsar Peter the Great founds St. Petersburg, the new capital city, on the northwest coast of Russia

1710 The Royal Chapel at Versailles, designed by Jules Hardouin Mansart (1645–1708), chief architect to French king Louis XIV (1638–1715), is completed

1710

1712 In the wake of strong pressure from Newton, Flamsteed's star catalogs—charting the position of nearly 3,000 stars—are published without his permission. Newton uses the data for his lunar theory

1715 The first performance of the *Water Music* by George Friederic Handel (1685–1759) takes place on the River Thames in London, England

1715 Louis XV (1710–1774), known as "Louis the Well-Beloved," is crowned king in France

1719 *Robinson Crusoe* by Daniel Defoe (1660–1731) is the first known English novel. Based on a true story, it describes the life of a man wrecked on a remote island

1720

1720 After Flamsteed's death, Halley becomes astronomer-royal at the Greenwich Observatory

1724 Tsar Peter the Great (1672–1725) founds the Academy of Sciences in Saint Petersburg, Russia

1726 English writer and clergyman Jonathan Swift (1667–1745) completes his satire on politics, *Gulliver's Travels.* Among the characters that Lemuel Gulliver meets are the inhabitants of Lilliput, who are only six inches high

1730

1732 American statesman and scientist Benjamin Franklin (1706–1790) begins publication of *Poor Richard's Almanac,* which gains a vast circulation in the American colonies

After 1730

1748 Swiss mathematician Leonhard Euler (1707–1783) publishes *Introduction to Infinitesimal Analysis;* later produces textbooks on differential and integral calculus

1905, 1915 German-American physicist Albert Einstein (1879–1955) presents his special and general theories of relativity (1905 and 1915), in which he revises aspects of Newton's laws on motion and gravity

CAROLUS LINNAEUS

1707–1778

"I don't believe that since the time of Conrad Gesner there was a man so learn'd in all parts of natural history as he...."

JAN FREDRIK GRONOVIUS
(1690–1762)

LINNAEUS WAS ONE OF THE GREATEST OF ALL NATURALISTS. HIS PRINCIPAL CONTRIBUTION TO SCIENCE WAS THE SYSTEM FOR CLASSIFYING PLANTS AND ANIMALS USING JUST TWO NAMES TO IDENTIFY EACH SPECIES. THIS SYSTEM, CALLED "BINOMIAL NOMENCLATURE," HAS REMAINED IN USE EVER SINCE.

CARL LINNAEUS WAS BORN ON MAY 23, 1707, IN Råshult, in southern Sweden, where his father was a clergyman. For someone who would spend a lifetime naming plants, the story behind his own name is an interesting one. Carl's father, Nils Ingemarsson Linnaeus, was from peasant stock and was simply called Nils Ingemarsson (Nils the son of Ingemar), as was customary in Scandinavia at the time. But when Nils enrolled as a student at Lund University, he had been obliged to add a formal surname in order to register. Names derived from Latin were fashionable in academic circles, so he called himself Linnaeus after a very fine small-leaved lime tree ("linn" in the local dialect). When his son, Carl, came to publish his work, he took the Latinization a stage further, calling himself Carolus Linnaeus.

Like many country parsons, Nils was an amateur botanist (botany is the study of plants), and it was through him that Carl acquired his enthusiasm for the subject. By the time he was eight years old Carl Linnaeus was already nicknamed "the little botanist."

A POOR PUPIL

In 1714, aged seven, Carl started at the high school in the country town of Växjö. His parents hoped that he would become a clergyman like his father; at school he studied Latin, Greek, theology, ethics, mathematics, physics, and logic. Linnaeus proved a poor student, but he did become good at Latin.

Johan Rothman, a local doctor and teacher, had noticed Linnaeus's interest in plants and gave him private lessons in medicine and botany. Among other topics, he taught him about sexual reproduction in plants. Rothman encouraged Linnaeus to abandon theology and study medicine instead. In those days many medicines were derived from herbs, so a good working knowledge of botany was

In a contemporary painting by the outstanding Dutch decorative artist Jacob de Wit (1695–1754), a group of characters examines a copy of Linnaeus's *Hortus Cliffortianus*, published in 1736.

a useful asset. In 1727 Linnaeus enrolled as a medical student at the University of Lund, transferring the following year to the University of Uppsala.

Uppsala Botanical Garden

Uppsala University had a botanical garden, to which Linnaeus was soon drawn. It was there that he met Olof Celsius, dean of Uppsala Cathedral and uncle of Anders Celsius (1701–1744), who devised the temperature scale that bears his name. Celsius, also a botanist, was impressed with Linnaeus's knowledge of plants, and introduced him to the garden's director, Olof Rudbeck Jr., who

Key Dates

1707	Born in Råshult, Sweden
1714	Starts at Växjö high school
1727	Attends the University of Lund
1728	Begins studying medicine at the University of Uppsala
1730	Becomes a lecturer in botany at Uppsala
1732	Mounts expedition to Lapland
1734	Travels to Falun; meets Sara Elisabeth Moraea
1735	Leaves for the Netherlands; graduates as doctor of medicine; publishes *Systema Naturae*, his classification of plants based on their sexual parts
1737	Publishes *Genera Plantarum*
1738	Returns to Sweden
1739	Appointed as first president of Royal Swedish Academy of Sciences; becomes physician to the Admiralty; marries Sara Elisabeth Moraeus
1742	Becomes professor of botany at University of Uppsala
1753	Publishes *Species Plantarum*, listing all known species of plants according to his system of classification
1761	Ennobled and changes his name to Carl von Linné
1774	Suffers a stroke
1778	Dies January 10

Linnaeus visited Lapland in 1732, where he acquired full Lapp costume, complete with bearskin gloves and a "magic" drum. He wore it for this 1737 portrait by Martinus Hoffman.

was an elderly man, and needed someone to take over from him. Rudbeck recognized Linnaeus's talent for the subject. Even though Linnaeus only in the second year of his studies, the director invited him to become a lecturer in botany. Linnaeus was especially interested in the structure of flowers. Developing the ideas he had acquired from Rothman, he became increasingly persuaded that it would be possible to introduce a new, much improved system for classifying plants based on their reproductive structures.

EXPEDITION TO LAPLAND

In 1732 Linnaeus set off on an expedition to the wilderness of Lapland, a region inhabited by the Lapp people and extending over the northern parts of Norway, Sweden, Finland, and Russia. He spent four months there, from May until September, journeying thousands of miles by foot to the Arctic Ocean, and discovering about 100 new species of plants.

His account of the expedition, *Flora Lapponica,* was published in 1737. Still interested in the reproductive structures of plants, it was in this report that he used the alchemical symbols ♀ for Venus and copper and ♂ for Mars and iron to signify "female" and "male" respectively.

A NEW LOVE

Linnaeus was interested in every aspect of natural history, and in 1733 lectured on mineralogy (the study of minerals) at Uppsala. In 1734, he traveled northwest of Uppsala to the county of Dalarna, an important mining center, to visit a copper mine at the capital, Falun. There he met a local doctor, Johan Moraeus, and his daughter Sara Elisabeth. Love blossomed very quickly, and two weeks after their first meeting he and Sara became engaged, though they did not marry until 1739.

Moraeus had qualified as a doctor in the Netherlands, and he persuaded Linnaeus to do the same. In June 1735 Linnaeus graduated as a doctor of medicine from the University of Harderwijk. He then moved to Leiden, also in the Netherlands. It was while he was there that he showed one of his manuscripts to the botanist Jan Fredrik Gronovius (1690–1762). Gronovius was so impressed that he published the work, *Systema Naturae,* at his own expense. It contained a classified list of plants, animals, and minerals and aroused much interest among Linnaeus's fellow naturalists.

Botanical Gardens

The botanical garden at Uppsala University was one of hundreds developed from the 16th century onward. The Renaissance that began in Italy in the 14th century brought a rebirth of interest in science, emphasizing the importance of very careful observation and recording; by the early 16th century European botanists were working with skilled illustrators to produce highly detailed books on plants, known as "herbals." In 1530, the German priest and botanist Otto Brunfels (c. 1488–1534) produced the first volume of his *Living Illustrations of Plants,* which marked the beginning of a more scientific approach to botany.

The interest generated by herbals in turn inspired universities in Europe to develop botanical gardens. The earliest, laid out in 1545, were at the universities of Padua and Pisa in Italy. Unlike most modern botanical gardens, which have plants for both study and ornament, these early botanical gardens derived from the medieval "physic" gardens (physic is an old word meaning medicinal), and chiefly contained herbs and other plants used in healing. University professors of

The botanical garden at the University of Padua was the first ever to be established in Europe. By the end of the 18th century there were estimated to be 1,600 such gardens in Europe.

Linnaeus continued to travel around Europe, visiting England before returning to the Netherlands. In 1737 he published *Genera Plantarum*, in which he expanded on his classification system for plants. The following year he visited Paris before finally returning to Sweden. On his travels he met and discussed his work with many leading botanists of the day, including the German Johann Dillenius (1687–1741), first professor of botany at Oxford University in England, and the French botanists, brothers Antoine and Bernard Jussieu (1686–1758 and 1699–1777 respectively).

PHYSICIAN AND SCIENTIST

Still just 31 years old, Linnaeus set himself up in practice as a physician in Stockholm. He was a successful doctor, and was widely admired too for his scientific work. In 1739 he became a founding member and first president of the Royal Swedish Academy of Sciences, and was appointed physician to the Admiralty.

In 1741 Linnaeus was appointed professor of practical medicine at the University of Uppsala, but within a year he had exchanged this for his preferred post: professor of botany.

medicine used these new botanical gardens both as an aid to teaching students and as a source of ingredients for making medicines.

As the science of botany became more established, botanical gardens were increasingly run by important botanists. In 1587 Charles de Lécluse, better known as Carolus Clusius (1525–1609), set up a collection of flowering bulbs at the University of Leiden; from this the Dutch bulb industry was developed. Just over 150 years later, in 1742, Linnaeus took over the supervision of Uppsala University's botanical gardens, which eventually contained about 3,000 species of plants.

Linnaeus made Uppsala the center of the world for botany. He also built a museum just outside Uppsala to house his huge collection of specimens. Today the botanical garden, Linnaeus's home, and the museum are all open to the public.

A 15th-century illustration shows a medieval doctor selecting herbs to use in medical treatments. Physic gardens were often placed next to the infirmary in medieval monasteries so that medicinal herbs could be collected easily.

Biological Classification

People had always given names to the plants that they saw around them, especially those that were used for food or fiber. This system worked adequately at first, but as naturalists studied more and more plants, and recognized more types of the same plant, their descriptions of them had to become more specific. So a certain type of buttercup, for example, had to be described as a "low-growing buttercup," and a certain type of low-growing buttercup had to be more specifically identified as a "low-growing buttercup with rounded leaves." In 1623 the Swiss anatomist and herbalist Caspar Bauhin (1550–1624) described one plant as a "low-growing, round-leaved, alpine buttercup with a smaller flower," writing it in Latin as *Ranunculus alpinus humilis rotundifolius flore minore*.

A Simpler Scheme

Latin names with six or more words were obviously difficult to remember. The scheme devised by Linnaeus was much simpler, and it is essentially the system we still use today. It is called "binomial nomenclature," which literally means assigning a name using two terms. The idea was not original. We often use two names to describe things, as in "table knife" or "bald eagle." Linnaeus, however, made the system more accurate and consistent.

Linnaeus gave each species a "specific" or "trivial" name. He then looked for species that resembled one another and sorted them into groups called genera (singular: genus, from the Latin word for race). The alpine buttercup described by Bauhin, for example, belonged to the genus called *Ranunculus*, and was given the specific name of *alpestris*. So its full name was *Ranunculus alpestris*.

Just two words identified the plant uniquely. Any other type of buttercup would be given a different specific name, so the two could never be confused. By convention, generic and specific names are printed in italic, and the generic name has a capital initial letter. Often, the name is followed by a letter or abbreviation referring to the person who made the name up. The alpine buttercup is *Ranunculus alpestris* L., the "L" standing for Linnaeus.

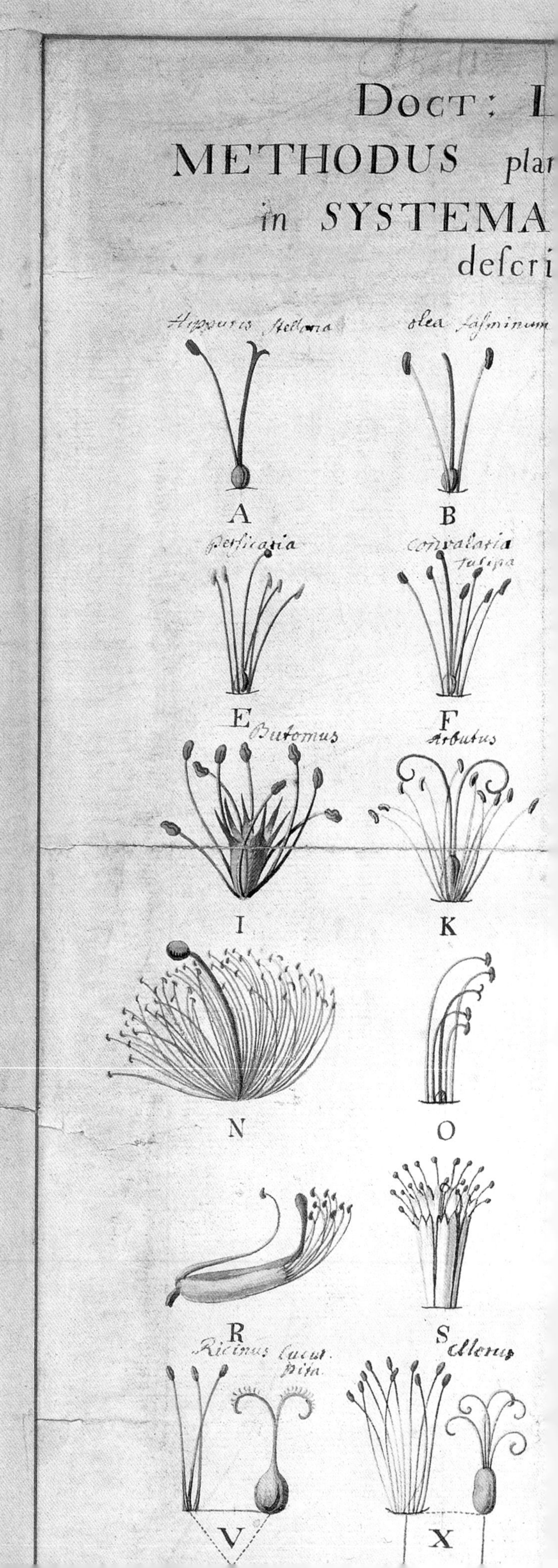

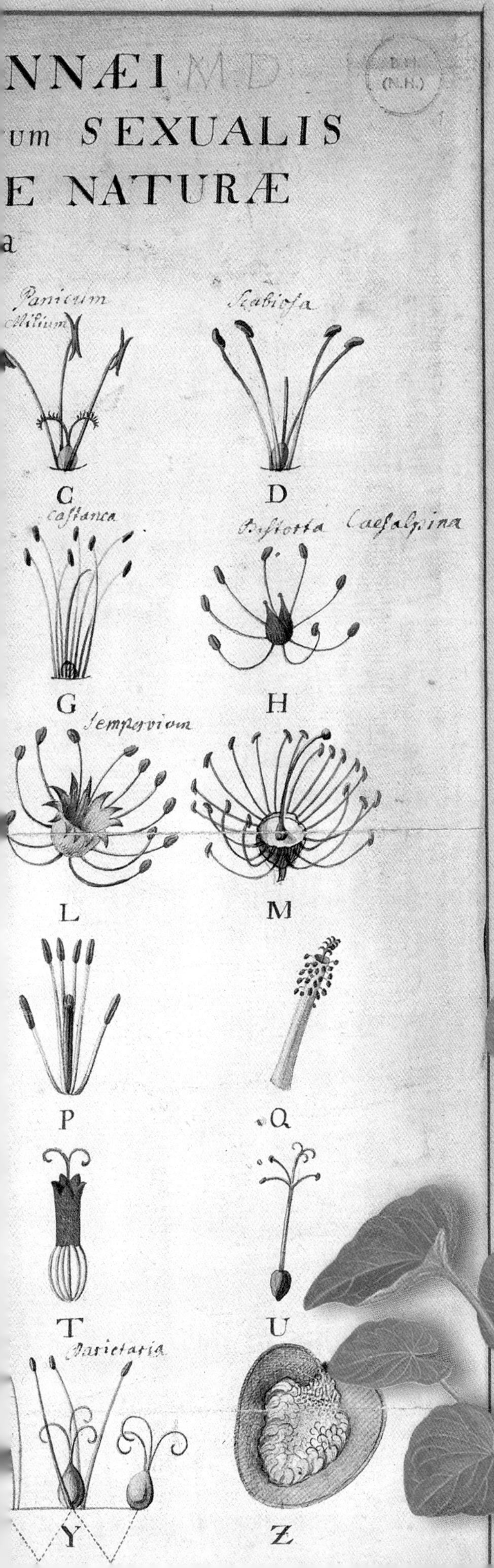

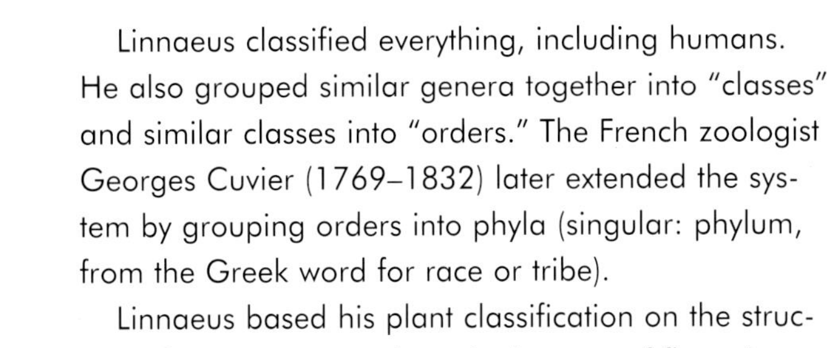

Linnaeus classified everything, including humans. He also grouped similar genera together into "classes" and similar classes into "orders." The French zoologist Georges Cuvier (1769–1832) later extended the system by grouping orders into phyla (singular: phylum, from the Greek word for race or tribe).

Linnaeus based his plant classification on the structures they use to reproduce; in the case of flowering plants he sorted them according to the number of stamens (the male reproductive part of a flower) and pistils (the female reproductive part) in their flowers. This system had flaws, as Linnaeus knew; plants that display the same number of stamens and pistils are not always related to each other, and modern classification is based on more complicated relationships between species, genera, and orders. Yet Linnaeus's system was simple to use and soon became popular. Now botanists everywhere, whatever language they spoke, knew exactly which plant was being described.

Classification Today

The naming of species is now regulated by strict rules set out in the *International Code of Botanical Nomenclature* and the *International Code of Zoological Nomenclature*. The rules apply to names at every level in the hierarchy, or graded order, of classification. From most specific to most general, this is traditionally made up of: species, genus, family, order, class, phylum (or, for plants, division), and kingdom. In the 1980s classification came to be based on systems that compared certain characteristics of species to see if they came from a shared ancestor or had evolved more recently. Biologists now class species by comparing their genes (the hereditary factors that pass from one generation to the next), leading to a further major rearrangement.

A page from Linnaeus's Systema Naturae *of 1735 shows his "sexual system" of plant classification, based on the number of flower parts.*

Preserving the Past

The Linnean Society was founded in 1788, and is the world's oldest biological society. Its first president was Sir James Edward Smith (1759–1828), an English medical student and naturalist. In 1783, aged just 24, Smith bought Linnaeus's vast collection of manuscripts and specimens from Linnaeus's widow in Sweden; she needed the money to provide dowries for her four daughters. Smith initially rented rooms in London in which he displayed the collection to a curious public.

The first meeting of the Linnean Society was held at Smith's home on April 8, 1788, its aim "the cultivation of the science of Natural History." The society remains a key forum for discussion on many aspects of science. In 1829, after Smith's death, the society bought the Linnean and the Smith collections and libraries, though the purchase plunged it into debt for some 30 years. More than 40,000 specimens from Linnaeus's collection are preserved in its London headquarters, alongside several other major plant and animal collections. The society also houses an important library specializing in works on the classification system that Linnaeus so successfully standardized.

CAROLI a LINNÉ,
Equitis Aur. de Stella Polari,
ARCHIATRI REGII, MED. & BOTAN. PROFESS. UPSAL.,
ACAD. PARIS. UPSAL. HOLMENS. PETROPOL.
BEROL. IMPER. LOND. ANGL. MONSPEL.
TOLOS. FLORENT. EDINB. BERN. &c.

SYSTEMA
NATURÆ
PER
REGNA TRIA NATURÆ,
SECUNDUM
CLASSES, ORDINES,
GENERA, SPECIES,
CUM
CHARACTERIBUS, DIFFERENTIIS,
SYNONYMIS, LOCIS.

TOMUS I.

EDITIO DUODECIMA, REFORMATA.
Cum Privilegio S:æ R:æ M:tis Sveciæ.

HOLMIÆ,
IMPENSIS DIRECT. LAURENTII SALVII,
1766.

The title page of Volume III of a 1768 edition of Systema Naturae, *and the lively frontispiece of* Hortus Cliffortianus *(1736). Linnaeus's many books, travel diaries, and manuscripts are preserved by the Linnean Society.*

Contribution to Science

Species Plantarum, published in 1753, is considered to be Linnaeus's most important work, listing all the species known at the time according to his system of classification. The great appeal of the system was that it was easy to use, enabling people to categorize species quickly. Together with the fifth edition of *Genera Plantarum* (1754), *Species Plantarum* remains to this day the starting point for botanical nomenclature of flowering plants and ferns.

Inspirational Teacher

Linnaeus was a prolific author, publishing about 180 books. To his students he stressed the importance of travel, urging them to visit every part of the world in search of new specimens. His talent as a teacher must have been considerable: 23 of his former students became professors. They spread the word of his work, as did Linnaeus himself through his many letters to the leading European naturalists of the day.

The king and queen of Sweden were among his patrons, and in 1761 he was made a nobleman. From then on he was known as Carl von Linné.

In 1772 Linnaeus's health began to fail and in 1774 he suffered a stroke. He died at Uppsala on 22 January, 1778, and was buried in the cathedral. His son succeeded him in his post as professor of botany at Uppsala University, and continued to add to his father's unique collection.

◀ **see also** Aristotle VOL 1:6
▶ **see also** von Humboldt VOL 2:24 Darwin VOL 2:54 Mendel VOL 2:86

LINNAEUS: Life and Times

SCIENTIFIC BACKGROUND

Before 1700

Aristotle (384–322 BC) makes a systematic study of plants and animals

English naturalist John Ray (1627–1705) classifies animal species into groups by their toes and teeth

Swiss botanist Caspar Bauhin (1560–1624) publishes a compendium of all known plants (*Pinax Theatri Botanici*)

1700

1720

1721 The German botanist Rudolph Camerarius (1665–1721) dies. Director of the botanic garden at Tübingen, he gained proof of sexuality in plants

1727 English botanist and chemist Stephen Hales (1677–1761) publishes *Vegetable Staticks*, on the physiology of vegetables

1730

1737 In *Genera Plantarum*, Linnaeus expands on his plant classification system

1738 *Ichthyology*, a systematic study of fishes by Linnaeus's friend and fellow Swede Peter Artedi (1705–1735), is published posthumously

1740

1742 Linnaeus becomes professor of botany at Uppsala University

1749 George-Louis Leclerc, Comte de Buffon (1707–1788) begins his 44-volume *Natural History*

1750

1753 Linnaeus publishes *Species Plantarum*, in which he lists and classifies all known species of plants

1754 Swiss naturalist and philosopher Charles Etienne Bonnet (1720–1793) publishes his *Study on the Use of Plant Leaves*

1760

1770

1771 Banks (1743–1820) is appointed director of the botanic gardens at Kew, in London

1771 English naturalist Joseph Banks (1743–1820) Banks returns from his epic voyage to the southern hemisphere with Captain James Cook (1728–1779); he brings back 800 previously unknown species of plants

1780

1788 The first meeting of the Linnean Society takes place in London

1790

After 1790

1800–1812 French naturalist Georges Cuvier (1769–1832) extends Linnaeus's classification system

1858 English naturalist Charles Darwin (1809–1882) announces his theory of evolution at the Linnean Society

POLITICAL AND CULTURAL BACKGROUND

1700 The Great Northern War begins between Sweden and Russia; it lasts until 1721 when Russia gains Swedish lands in the Baltic

1703 Peter the Great (1672–1725), tsar of Russia, founds the city of St. Petersburg as his northern capital

1725 English writer Jonathan Swift (1667–1745) publishes *Gulliver's Travels*

1733 *Mass in B Minor*, the great choral work by German composer Johann Sebastian Bach (1685–1750) is performed for the first time

1742 George Friederic Handel (1685–1759) completes his oratorio, *The Messiah*, which receives its first performance in Dublin, Ireland

1751 French writer Denis Diderot (1713–1784) publishes the *Encyclopedia*, a key work of the Enlightenment

1755 Lisbon, capital of Portugal, is destroyed in a devastating earthquake

1762 Catherine the Great (1729–1796) becomes empress of Russia

1768–71 Captain James Cook makes his first Pacific voyage, discovering New Zealand and Australia

1774 The first Shakers colony is founded in the United States; the Christian group is an offshoot of the Quakers

1776 The American Declaration of Independence is signed on July 4

1788 *The Times* newspaper is founded in London

1789 The French Revolution begins

ANTOINE LAVOISIER

1743–1794

"It required only a moment to sever his head and probably one hundred years will not suffice to produce another like it."

JOSEPH-LOUIS LAGRANGE
French mathematician and deviser of the metric system, speaking of Lavoisier's death in 1794

ANTOINE LAURENT LAVOISIER IS OFTEN KNOWN AS THE "FATHER OF MODERN CHEMISTRY," MOST ESPECIALLY FOR HIS WORK IN EXPLAINING THE CHEMICAL BASIS OF COMBUSTION (HEAT OR BURNING). HIS REDEFINITION OF THE ELEMENTS FORMED THE BASIS FOR A NEW LANGUAGE OF CHEMISTRY, BUT HIS CAREER UNFOLDED DURING THE YEARS LEADING UP TO THE FRENCH REVOLUTION AND HIS LIFE ENDED, TRAGICALLY, ON THE GUILLOTINE, WHERE HE WAS EXECUTED BY THE REVOLUTIONARY FORCES.

ANTOINE LAURENT LAVOISIER WAS BORN IN PARIS on August 26, 1743, the son of a leading lawyer. He was educated in law at the best school in Paris, the Collège Mazarin, where chemistry, mathematics, astronomy, and botany were also taught. It was these subjects, rather than his legal studies, that fired the young scholar's imagination, and he soon began to devote his spare time to scientific experiments.

EARLY VENTURES

Lavoisier graduated in 1763 but did not join the legal profession as his father had intended. Instead, he accepted an invitation to help the scientist Jean-Etienne Guettard (1715–1786) carry out a geological survey of France. The task took three years. At the end of it Lavoisier wrote a paper on the properties of gypsum, a mineral also known as "plaster of Paris" because it was used to cover the walls of Parisian houses. He presented the paper at the foremost institution of learning in the country, the Academy of Sciences.

The next year, 1766, Lavoisier entered an Academy competition to find a better system of street lighting for Paris. At this time, the only form of lighting in towns was with oil lamps or candles. The development of coal gas or electric lighting lay many decades ahead. Lavoisier's essay, on various forms of lighting apparatus, did not win him the prize, but its treatment of the subject so impressed the judges that King Louis XV (1710–1774) ordered him to be given a special medal. Lavoisier's name was already becoming known.

KEY DATES

1743	Born in Paris, France
1765	Publishes paper on gypsum
1768	Joins a private tax-collecting agency
1770	Proves that water cannot be turned into earth
1772–75	Shows that metals burn by absorbing "vital principle" from air
1777	Identifies "dephlogisticated air" as oxygen
1783	Shows that water is a compound of hydrogen and oxygen
1785	Becomes director of the Academy of Sciences
1787	Defines concept of the chemical element and publishes *Method of Chemical Nomenclature*
1789	Publishes *Elementary Treatise of Chemistry*
1794	Condemned by revolutionary tribunal and guillotined

In this contemporary etching, Lavoisier is pictured in his laboratory. In 1768 the Academy of Sciences in Paris tried to find out whether water supplied to Paris through an open canal was fit to drink. This started Lavoisier on a series of experiments on water; he is shown here explaining his results to two observers.

Lavoisier had a large inheritance from his wealthy family but still needed a steady income to fund his scientific studies. So, in 1768 (the same year he was elected to the Academy), he bought a stake in the General Farm (*Ferme Générale*), a tax-collecting agency used by the royal government to collect taxes on tobacco, salt, and other goods. Agency members, known as tax farmers, made great profits for themselves and were universally disliked. Their activities were a chief cause of grievance against the French monarchy in the years leading up to the Revolution. Joining the Farm would prove to be a fateful decision for Lavoisier.

Meanwhile, in 1771, at the age of 28, he married Marie-Anne Paulze, the 14-year-old daughter of a fellow member of the agency. Despite the great age difference, it was a very happy marriage. Marie-Anne shared her husband's interests and helped him in his laboratory experiments. Lavoisier's private fortune enabled him to purchase the best scientific equipment that money could buy. He placed his talents at the disposal of the Academy of Sciences, preparing reports on a variety of subjects, ranging from early balloon flight and hypnotism to the manufacture of gunpowder, inks, and dyes, and the reasons why iron rusts.

WATER INTO EARTH?

In 1768 the Academy of Sciences decided to find out whether water supplied to Paris through an open canal was fit to drink. The usual way of testing for impurities was to boil water dry in order to see what solids were left. These investigations started Lavoisier on a new line of scientific inquiry: whether it was possible for one substance (water) to be changed into another (earth). According to the medieval alchemists—scholars who blended philosophy and mysticism with practical chemical experimentation—it was. The idea that some substances could be changed (or "transmuted")

Georg Stahl
1660–1734

Stahl, born in Ansbach, southern Germany, was a physician and chemist. After teaching at the University of Halle, he spent the final years of his life in Berlin as physician to the king of Prussia, Frederick William I (1688–1740). He is best remembered for his ideas on combustion, or burning. According to Stahl, when something burned, a substance called phlogiston was released into the air. He also believed that phlogiston was lost when metals corroded, or rusted. His ideas, which built on those of earlier scientists such as Johann Joachim Becher (1635–1682), dominated chemical thought for almost a century until disproved by Lavoisier. Stahl also believed in animism—the idea that all living organisms are guided by a "life force."

This painting by English artist Joseph Wright (1734–1797) depicts the traditional view of an alchemist's workshop. Alchemists sought to turn metals such as lead into gold, and much of their work was shrouded in secrecy. However, their investigations often involved complex chemical procedures and their laboratories were far from being the witches' kitchens so often portrayed.

Chemical Apparatus

Before the 18th century, preparing and collecting gases—an essential part of chemical experimentation—was a difficult and sometimes dangerous affair. The word "gas"—which was first introduced in the 16th century, though most scientists continued to use the term "air"—comes from the Greek word for chaos. These mysterious, invisible, but often pungent substances had a habit of behaving unpredictably and explosively during experiments.

As interest grew in studying the chemistry of gases, great improvements were made to chemical apparatus. To collect gas from a heated solid and remove any impurities, equipment was devised that bubbled gases through water before collection. The "pneumatic trough"—a shelf with a hole to support a flask that enabled gas samples to be transferred from one container to another—became a standard piece of apparatus. English chemist Joseph Priestley (1733–1804) refined the pneumatic trough and, replacing the water with mercury, used it to collect many gases that would have otherwise dissolved in water.

For his investigations of gases, however, Lavoisier often preferred to use a gasometer, a sophisticated and expensive apparatus of his own devising. The search for accuracy held the key to Lavoisier's revolution in chemistry. Much of his personal wealth was spent on inventing elaborate equipment for weighing and measuring the gases given off in chemical reactions. Such attention to precise measurement is known as "quantative analysis," and lies at the very heart of modern chemistry.

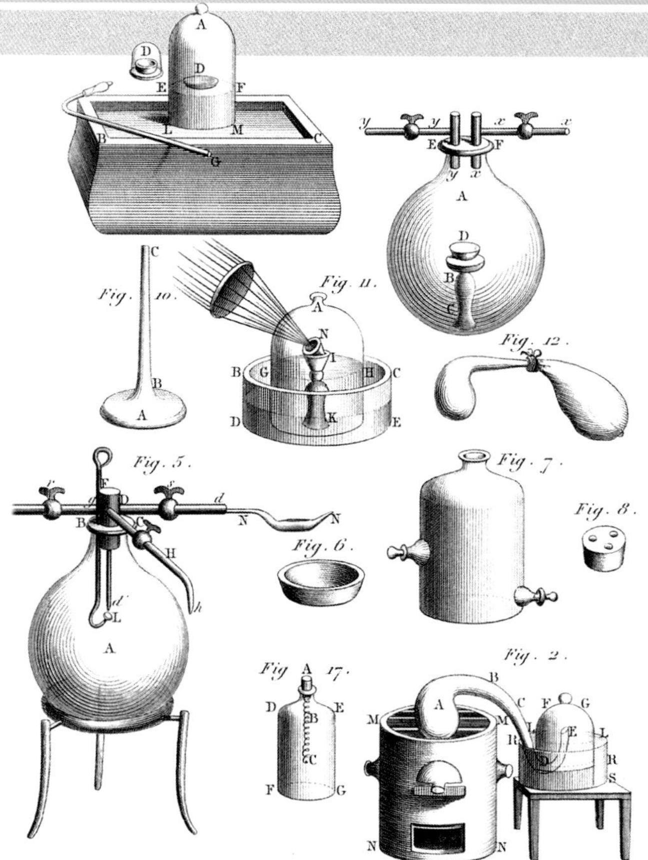

A plate from Lavoisier's An Elementary Treatise of Chemistry *(1787). More than a third of this groundbreaking textbook on chemistry was given over to descriptions and detailed illustrations of chemical apparatus, an indication of the importance Lavoisier attached to this aspect of his work.*

Lavoisier's wife, Anne-Marie, was a trained artist who prepared many of the illustrations in his published works. This drawing depicts an experiment on respiration (breathing) using a complex piece of apparatus. Lavoisier's assistant, seated center, breathes air into a mask, which passes through a retort (glass vessel) to remove certain substances. Lavoisier stands over a chamber that is collecting the remaining gases for analysis. Madame Lavoisier, seated far right at the table, is recording the results.

The Academy of Sciences

The French Academy of Sciences was established in Paris in 1666, four years after the foundation of the Royal Society of London (see box pages 62–63). It later received the royal approval of King Louis XIV (1638–1715), who saw it as a way of boosting France's scientific talent in the face of his rivals. When countries were frequently at war, on land and at sea, there were practical reasons for encouraging scientific advances in such areas as military technology (gunpowder and artillery), and navigation. Scientific research could also improve key national industries such as textiles and agriculture.

The Academy of Sciences originally covered six areas of science—mathematics, astronomy, mechanics, chemistry, botany, and anatomy. Membership was strictly limited to 18 working "academicians," paid by the government to advise it on scientific affairs and to report on official questions put to them as a body. There were also a number of honorary members, working associates and, right at the bottom, unpaid assistants. It was to this lowest position that Lavoisier was admitted in 1768. He rose through its ranks to become the Academy's director in 1785 and its treasurer in 1791.

Being made an academician was a great honor for any French scientist. But the structure of the Academy mirrored the rigid divisions of French society of the time. Honorary membership was only ever given to aristocrats, and members were elected or promoted on grounds of seniority rather than merit. Vacancies for election often led to intense lobbying, disputes, and ill-feeling.

One of the leaders of the French Revolution, Jean-Paul Marat (1743–1793), was an amateur scientist with a keen interest in optics and electricity. The Academy repeatedly rejected his attempts to gain admission in the 1780s. His failure to join its ranks was probably a factor in the revolutionary government's decision to close the Academy down in 1793. But no government could survive long without scientific advice, and two years later the Academy became part of a new National Institute.

Louis XIV visits the Academy of Sciences in its early years. In the distance the Paris Observatory, for studying astronomy, can be seen under construction.

into others when heated derived from Aristotle's theory of the four elements of water, earth, air, and fire (see page 12). It lay at the root of the alchemists' age-old search for the "philosopher's stone" that would turn metals such as lead, tin, or copper into gold or silver. But alchemy was not just a question of magic and mysticism. In many respects it was the forerunner of chemistry; no less a scientist than Isaac Newton is known to have carried out alchemical experiments (see page 67).

Even in the late 18th century some scientists were still prepared to argue that water turned into earth on heating. This was based on the fact that when pure rainwater was distilled (condensed) in a glass vessel, solid matter ("earth") was left behind. However, Lavoisier suspected that this "earth" was dissolved, or leached from, the glass itself during the boiling process. Using some of the most sophisticated measuring devices available, he weighed both the glass vessel and the water before and after heating, time after time, over a period of three months. He found that the weight of the residue of "earth" more or less equalled that lost by the vessel; as he had suspected, the "earth" had not come from the water, but had been produced from the glass apparatus itself.

> ***"Chemists have made phlogiston a vague principle... sometimes it has weight, sometimes it has not, sometimes it is free fire, sometimes it is fire combined with an earth.... It is a veritable Proteus that changes its form for every instant!"***
>
> LAVOISIER IN AN ESSAY ATTACKING THE THEORY OF PHLOGISTON (1785)

THE PHLOGISTON THEORY

Lavoisier's attention now turned toward an investigation of what takes place when substances are burned (combustion). According to German chemist Georg Stahl, all substances were ultimately made up of water and three varieties of earth, one of which was combustible. This, he argued, was set free into the air when the substance was burned. He therefore called it phlogiston, from the Greek word for "burned." According to Stahl, metals burned because they lost phlogiston into the atmosphere, whereas we now know that the reverse is true: they burn by uniting with oxygen in the atmosphere.

Lavoisier's interest in combustion first developed in 1771 when the amateur chemist Louis-Bernard Guyton de Morveau (1737–1816) showed that metals grew heavier when they were roasted in air. Guyton, in common with all scientists at that time, believed in Stahl's phlogiston theory. If phlogiston was lost during combustion, metals would lose weight. In order to make the facts fit with the theory, Guyton claimed that metals became heavier when "dephlogisticated" because phlogiston was a weightless substance that buoyed up the materials containing it. Lavoisier seriously doubted this conclusion. He thought that a more likely explanation was that "air" (the term then used for a gas) was somehow involved in the process of combustion, and that this air caused the increase in weight.

THE MYSTERIES OF COMBUSTION

Lavoisier burned a variety of substances in order to measure the changes in weight that took place. He reported his findings in November 1772, and concluded that phosphorus and sulfur increased in weight when burned because they absorbed air. He then reversed the process by roasting a burned metallic residue with charcoal. Scientists at the time called the crumbly residue left after a metal has been burned or roasted (calcinated) the calx. We now know it as an oxide (a compound formed when any substance mixes with oxygen). The metal after roasting weighed less than the original calx, suggesting to Lavoisier that it had lost the air absorbed in the original burning. Lavoisier used litharge, which is an oxide of lead, for these experiments.

Lavoisier now became aware that a group of chemists then at work in Britain had already discovered that the atmosphere is made up of several different "airs." These chemists are known as the "pneumatic chemists" from *pneuma*, the Greek word for "air." For example, Joseph Black (1728–1799), working in Scotland in the 1750s, had shown that some substances (those we now refer to as "carbonates") contained an air (now known as carbon dioxide) that was heavier than ordinary atmospheric air and did not combust (burn). Soon afterwards, the English chemist Henry Cavendish (1731–1810) found that a very light, inflammable air (one that burned easily) was produced when a solution of water and sulfuric acid was poured over iron. Cavendish, a firm believer in the phlogiston theory, thought he had isolated pure phlogiston.

Keenly aware of his own ignorance, Lavoisier devoted the whole of 1773 to studying the history of chemistry. He read up on everything that had been written previously about air, or airs, and repeated the experiments of the pneumatic chemists. At first this only succeeded in confusing him. For instance, he came to the conclusion that carbon dioxide (then known as "fixed air") in the atmosphere was responsible for the burning of metals and the increase in their weight.

In 1774 the English chemist Joseph Priestley (1733–1804) made an important discovery. Heating oxide of mercury in a closed vessel, he collected a gas ("dephlogisticated air") in which things burned far better than they did in ordinary air. Priestley reported the results of his experiments directly to Lavoisier when he made a visit to Paris that same year. As a result of this momentous meeting, Lavoisier carried out a series of experiments of his own. These gradually led him to the realization that this "healthiest and purest part of air," as he called it, was in fact the active agent of burning, the key that unlocked the mystery of combustion.

Henry Cavendish
1731–1810

One of the most gifted and original scientists of his day, Henry Cavendish made important discoveries in many fields, though his work went largely unrecognized in his own lifetime. A member of the aristocracy, he inherited a fortune from his uncle which enabled him to buy books and equipment for his scientific investigations. But he was a recluse who avoided contact with other people and published very little.

One of the so-called British "pneumatic chemists", his specialty was the study of gases released from solids and liquids. His experiments led him to isolate the gas that came to be known as hydrogen and his work also helped later chemists discover other gases, such as argon and neon. In physics, he carried out pioneering research into electricity.

The Cavendish Physical Laboratory of Cambridge University, England, famous as a center of research into radioactivity in the early 20th century, was named for him and has a museum of his apparatus.

Marie-Anne Lavoisier
1758–1836

Although Marie-Anne Pierrette Paulze, whom Lavoisier married in 1771, was not an original scientist herself, she made an important contribution to science through the work she shared with her husband. She assisted her husband in his laboratory and, having studied art specially, prepared the illustrations in his *Treatise*. She also learned English in order to translate the works of Cavendish and Priestley for her husband to study. Her final service for Lavoisier was to collect and edit his *Memoirs of Chemistry* (1803).

In 1805 she married the American-born scientist Count Rumford (1753–1814), known for his work on heat. The marriage was not a happy one and they separated in 1809. Rumford is reported to have commented on the good fortune of Lavoisier to have been guillotined.

Marie-Anne Lavoisier studied art under the artist Jacques-Louis David (1748–1825), one of the greatest French painters of the day, who later became the official artist to the emperor Napoleon Bonaparte (1769–1821). David painted this spectacular portrait of Lavoisier with Anne-Marie standing lovingly at his shoulder in 1788.

PHLOGISTON VERSUS OXYGEN

Because the new part of air that Priestley had discovered burned carbon to form the weak acid, carbon dioxide, Lavoisier gave it the name "oxygen," which literally means "acid-former" (from the Greek *oxys*, meaning acid). He now correctly asserted that combustion occurs when oxygen combines with another substance, releasing (or evolving) heat and light, and causing that substance to increase in weight.

Lavoisier had prepared the ground for a totally new theory of chemistry in which phlogiston played no part. But it was some years before he spelled out his new theory in full. The chief reason for this was that it remained difficult, without the phlogiston theory, to explain why an inflammable gas was produced when dilute acid was poured on metal. Strangely enough, Lavoisier arrived at the answer to this puzzle through a finding made by Henry Cavendish, the original discoverer of the inflammable gas (or phlogiston as he firmly believed it to be).

Repeating an earlier experiment of Joseph Priestley's in 1783–4, Cavendish found that when a mixture of oxygen and the inflammable gas were exploded by means of an electric spark, moisture covered the sides of the vessel. Cavendish concluded that this was water. To Lavoisier, the experiment was clear evidence that water is made up of parts of oxygen and of Cavendish's inflammable gas. Assisted by Pierre Simon Laplace (1749–1827), he demonstrated that when oxygen and the inflammable gas were burned together in a closed vessel, water was formed. Therefore Lavoisier gave this inflammable gas the name "hydrogen," meaning "water-forming." He could now explain why hydrogen was given off when metal was dissolved in dilute acid. It came not from the metal itself (as the phlogistonists claimed) but from the water in the dilute acid as it was broken down into its parts of oxygen and hydrogen.

THE NEW CHEMISTRY

Lavoisier's insight revolutionized chemistry (the science that deals with the composition of substances, and the changes they undergo). Since the time of the Greeks, people had believed that all matter was made up of one of four elements (earth, air, water, and fire). Lavoisier now rewrote the

Joseph Priestley
1733–1804

Joseph Priestley was a man of many interests and talents who in his lifetime won fame (and some notoriety) as a Nonconformist clergyman, teacher, and political radical. In science he is best remembered as a pioneer of chemistry.

Priestley's introduction to science came in the 1760s, when he was inspired by the American scientist Benjamin Franklin (1706–1790) to begin research into electricity. However, he soon turned to chemistry, especially the study of gases. One of his earliest achievements was to identify the properties of carbon dioxide. He found that it was heavier than air, could put out flames, and, when dissolved in water, produced a refreshing, fizzy drink ("soda water") ideal for long sea voyages.

In a period of five years, he discovered 10 new gases. His preparation of oxygen was crucial in leading Lavoisier to a true understanding of combustion, yet he refused to abandon the phlogiston theory himself.

His ardent support for the French Revolution caused great hostility in England. After his home and laboratory were burned down by a mob, Priestley spent the rest of his life in America.

A disrespectful English cartoon mocks the French general Napoleon Bonaparte, later to become emperor of France. He is taking part in an experiment by "the great chemist Lavoisier in Paris," with stupendously explosive results.

description of elements, defining them as any substance that cannot be broken down into other substances but out of which all other substances are formed. Hydrogen and oxygen are both elements. They cannot be broken down into simpler substances; nor can they be made. Water, however, is a compound. It is made up of two elements, hydrogen and oxygen, as Lavoisier and Laplace had demonstrated.

By 1785 Lavoisier was confident enough to launch a full attack on the old chemistry. All chemical change could now be explained without resorting to the phlogiston theory. He built up a team of younger assistants around him and held twice-weekly discussions and demonstrations of his latest findings at his home. A new language was needed to describe the processes of chemical change, and Lavoisier and his followers founded a journal in 1788 to put across their ideas, *Annals of Chemistry*, which still exists today. In 1789 Lavoisier published *An Elementary Treatise of Chemistry*. Written in clear, logical language, it became the standard textbook on chemistry for many decades.

How Animals Breathe

Lavoisier and his assistants carried out a series of experiments that confirmed that animal respiration (breathing) is a slow form of combustion. The oxygen that a living being breathes in burns the carbon in foodstuffs. This

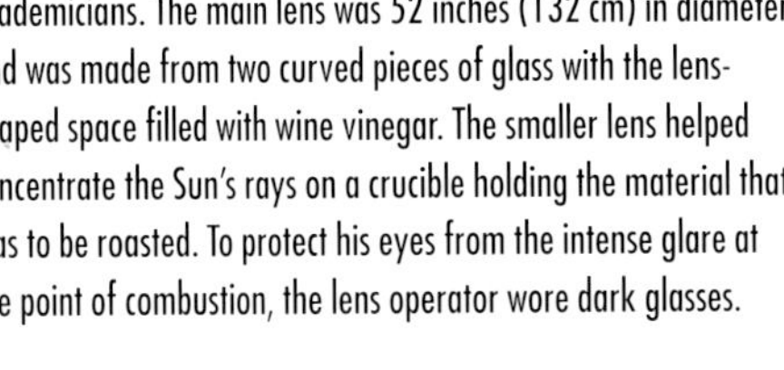

The great burning lens that was made for the Academy of Sciences in 1774 under the supervision of Lavoisier and other academicians. The main lens was 52 inches (132 cm) in diameter and was made from two curved pieces of glass with the lens-shaped space filled with wine vinegar. The smaller lens helped concentrate the Sun's rays on a crucible holding the material that was to be roasted. To protect his eyes from the intense glare at the point of combustion, the lens operator wore dark glasses.

The New Language of Chemistry

In 1782 French scientist Guyton de Morveau (1737–1816), inspired by Swedish naturalist Carolus Linnaeus's system for the naming of plants (see box pages 74–75), proposed that similar rules should be drawn up for naming chemicals. Until now, chemicals were referred to by a variety of flowery terms derived from alchemy and pharmacy. Sulfur and mercury, for example, were known as "father and mother," ammonia as "aquila coelestis" or "heavenly eagle." Chemists and pharmacists also used obscure symbols, derived from ancient sources, for denoting substances in chemical formulae or medical prescriptions. There was no common system that everyone could understand.

Lavoisier realized that his new chemistry provided the basis for a precise, easily understood naming system, and from 1787 he worked with Guyton de Morveau and two other colleagues to produce a 300-page dictionary of chemical names. They started by drawing up a list of 55 elements (substances that cannot be broken down into further parts). These names were the building blocks of their naming system, and chemical compounds were given names that clearly indicated the elements that combined to form them. So "oil of vitriol" became "sulfuric acid" and "flowers of zinc" became "zinc oxide." Lavoisier's new language of chemistry is still used today, with many additions to the original list.

A "tree of universal matter" from an alchemical text. Alchemical knowledge retained an aura of mystery and used obscure symbols and names. Lavoisier replaced this with a practical language of chemistry that could be easily understood by everyone.

chemical reaction releases heat (a vital form of energy for the animal's survival), and also carbon dioxide, which is breathed out as a waste product into the atmosphere. Lavoisier and Laplace demonstrated these processes in a series of experiments using guinea pigs. This is said to be the origin of the expression "to be a guinea pig," applied to anyone who is the subject of scientific testing. Lavoisier's work on animal respiration laid the foundations of biochemistry (the science that deals with the chemical processes of living organisms).

Execution of a Scientist

In July, 1789 the French Revolution broke out. Anger against the king and the privileged classes had been mounting for years. The government was bankrupt, taxes were higher than ever, and food shortages were widespread, sowing the seeds of rebellion. The overthrow of the old order and the election of a new National Assembly brought about a number of popular reforms.

Lavoisier supported the early stages of the Revolution, sitting as a political deputy. He wrote a major review of France's finances and agricultural resources, undertook research into the quality of gunpowder, and was part of a committee set up by the Academy of Sciences to create a uniform system of weights and measures.

Lavoisier was arrested on November 24, 1793 and taken before a revolutionary tribunal five months later. The revolutionary leader and failed Academician Jean-Paul Marat (left) led the accusations against him.

But the initial unity of the Revolution soon gave way to political division and bitter quarrels. A prime target for hostility were former members of the royal tax-collecting agency, the General Farm, of whom Lavoisier was one. To make matters worse, he was on record as having proposed the building of a wall around Paris to curb smuggling only a few years before the Revolution.

In November 1793 Lavoisier was arrested, along with other members of the Farm. He might have escaped the death penalty, had it not been claimed at his trial that he had been corresponding with France's political enemies abroad. It was useless to point out that these letters were on scientific matters. He was sentenced to death and executed that same day, May 8, 1794. It is said that when he asked for a delay in his execution to finish an important piece of research, he received the curt reply: "The Republic has no need of experts." In a last letter to his wife he wrote: "I have had a very happy life, and I think I shall be remembered with some regrets and perhaps leave some reputation behind me. What more could I ask?"

Science and the French Revolution

Although the head of the tribunal that tried Lavoisier for treason in 1794 is reputed to have declared publicly that, "The Republic has no need of experts," this clearly was not true. The leaders of the Revolution strongly believed that the new society they were creating was founded on principles of reason and natural justice. Great faith was placed in science to improve the life of the common citizen.

The Metric System

Along with the creation of a new political system came the opportunity to introduce a standardized system of weights and measures to replace the confusion of different units in use before the Revolution. Before its suppression by Marat in 1793, the Academy of Sciences had set up a commission, whose members included Lavoisier, to look at the subject. It was decided to base the new system on multiples of 10. The standard unit of measurement was the meter (from the Greek *metron*, meaning "measure"), one ten-millionth of the distance between the North Pole and the Equator. The gram become the standard unit of weight, and the liter the standard unit of volume. This system is internationally recognized by scientists, and is used today in most countries of the world.

The introduction of a metric system of weights and measures not only had clear benefits for scientific research, as foreseen by Lavoisier, but it also improved the sale of goods in shops and commercial businesses. Standard measures were clearly marked, and customers could ensure that the shopkeeper was not cheating them. Weights could be inspected regularly by the authorities to stop them being tampered with.

▶ **see also** Pasteur VOL 2:76 Mendeleev VOL 3:6

The revolutionary enthusiasm for metrication spread to the calendar. The 7-day week was abandoned and each 30-day month was divided into three periods of 10 days, the last day of each being a rest day. The months were given new descriptive names such as Floréal ("blossom") for April 20 to May 19 and Brumaire ("mist") for October 22 to November 20. However, the new calendar never became popular and was eventually scrapped by Napoleon in 1805.

Gunpowder from Manure

Revolutionary France was soon at war with its neighbors, who were terrified that revolutionary ideas would spread to unsettle their own governments. French ports were blockaded to prevent essential supplies from entering the country, halting the importation of saltpeter, an essential ingredient in making gunpowder. An ingenious technique was developed for converting the calcium nitrates present in farmyard manure into saltpeter by heating them with soda (sodium carbonate). Before the Revolution a chemist, Nicolas Leblanc (1742–1806), had found a way of making soda (also used in soap and glass manuacture) in quantity based on salt and sulfuric acid. In 1791 he was granted a patent for his invention and a factory was built to produce soda. It was confiscated by the revolutionary government in 1793, though later returned to Leblanc by the emperor Napoleon in 1802. Leblanc's process, which was environmentally highly polluting, later formed the basis of the heavy chemical industry in 19th-century Europe.

Scientific Education

One of the most enduring impacts of the Revolution was the importance it gave to scientific education. The Polytechnic School, founded in 1794 with an emphasis on mathematics, engineering, and chemistry, furnished most of the next generation of research chemists. Napoleon also did much to boost scientific and industrial development. To ensure a supply of scientists for the future, the "lycée" system of secondary education—still in existence today—was started.

The storming of the Bastille prison in Paris on July 14, 1789, began the French Revolution.

LAVOISIER: Life and Times

SCIENTIFIC BACKGROUND

Before 1760

Irish physicist and chemist Robert Boyle (1627–1691) and English chemist John Mayow (1640–1679) experiment on air, respiration, and combustion; German chemist Johann Joachim Becher (1635–1682) argues that combustion is due to sulfurous earth

German chemist Georg Stahl (1660–1734) develops phlogiston theory

English chemist Stephen Hales (1677–1761) shows that air is involved in chemical reactions

1760

1763–66 Lavoisier makes a geological tour of France with Jean-Etienne Guettard (1715–1786)

1765 Lavoisier publishes his first scientific paper, on gypsum

1765

1766 English chemist Henry Cavendish (1731–1810) prepares inflammable air (hydrogen)

1768–70 Lavoisier proves that water cannot be changed into earth

1768 Lavoisier is elected to the Academy of Sciences in Paris

1770

1771–72 French chemist Baron Louis Bernard Guyton de Morveau (1737–1816) shows that metals gain weight when burned in air

1771–72 Swedish chemist Carl Wilhelm Scheele (1742–1786) prepares what he calls "fire air," actually oxygen

1772–75 Lavoisier shows that metals burn by absorbing "vital principle" from air

1774 Lavoisier publishes *Opuscles physiques et chimiques*, summarizing his experimentation

1774 English chemist Joseph Priestley (1733–1804) prepares "dephlogisticated air" (oxygen)

POLITICAL AND CULTURAL BACKGROUND

1751 The first volume of the *Encyclopedia*, edited by Denis Diderot (1713–1784), is published; it will be a landmark of the Enlightenment in France

1760 France surrenders Montreal, Canada, to Britain

1762 French philosopher and writer Jean Jacques Rousseau (1712–1778) produces *The Social Contract*; his demand for "Liberty, Equality, and Fraternity" becomes the rallying cry of the French revolutionaries

1768–71 English navigator James Cook (1728–1779) undertakes his first Pacific voyage aboard the *Endeavor*; he is the first European to chart New Zealand and the east coast of Australia

1773 Paul Revere (1735–1818) is among those who takes part in the "Boston Tea Party," the destruction of chests of tea at Boston Harbor to protest at the high duties imposed on tea by the British government

1774 Louis XVI, the last prerevolutionary king of France, comes to the throne

1775

1776 Lavoisier decides that all acids contain oxygen

1775 Priestley notices that "dew" is formed when hydrogen explodes with oxygen

1777 Lavoisier develops theory of gaseous state which involves heat as a principle of expansion; identifies "dephlogisticated air" as oxygen

1775 Thirteen of Britain's North American colonies rebel against the government of George III, heralding the beginning of the American War of Independence

1778 France intervenes on the side of the colonists against Britain in the American War of Independence

1780

1780 Scheele publishes his book on gases, *Chemical Observations on Air and Fire*

1782 Guyton de Morveau urges a reform of chemical language

1783 Lavoisier shows that water is a compound of hydrogen and oxygen

1784 Lavoisier works with Pierre Simon Laplace (1749–1827) on calorific heat theory and animal heat using guinea pigs

1780 The beginning of the Industrial Revolution in England is marked by the expansion of the cotton industry

1783 The Treaty of Versailles acknowledges the independence of the colonies and the establishment of the United States of America

1784 *Encyclopedia Britannica*, first published 1768–1771, appears in a new 10-volume edition

1785

1787 With a group of French chemists, Lavoisier publishes the influential book *Method of Chemical Nomenclature*, which classifies known elements and compounds

1789 Lavoisier publishes his textbook, *Elementary Treatise of Chemistry*

1789 The French Revolution begins with the storming of the Bastille, the royal prison in Paris which was a hated symbol of royal tyranny. It was found to contain only a handful of prisoners

1790

1793 The Academy of Sciences is suppressed as France's Reign of Terror takes hold

1794 Lavoisier is executed on May 8

1794 Priestley emigrates to America after mob attacks his house; he remains a convinced phlogistonist

1793 Louis XVI is tried and executed in Paris, one of more than 2,500 people to be guillotined in Paris during the Reign of Terror

1795, 1807 France captures large parts of Prussia (northern Germany and Poland)

After 1795

1798 Count Rumford (Benjamin Thompson) (1753–1814) concludes that heat is a kind of motion; he marries Lavoisier's widow in 1804

1802–04 English chemist John Dalton (1766–1844) develops an atomic theory based on Lavoisier's elements

1805 Lavoisier's *Chemical Memoirs* are published posthumously by his widow and collaborator, Marie-Anne Paulze

1804 French general Napoleon Bonaparte (1769–1821) crowns himself Emperor of France

Index

Page numbers in *italic* type refer to picture captions. **Bold** page numbers refer to the main discussion of the subject.

Picture credits

Abbreviations
AKG Archiv für Kunst und Geschichte, London
ARPL Ann Ronan Picture Library/Image Select
BAL Bridgeman Art Library
C Corbis
MEPL Mary Evans Picture Library
SPL Science Photo Library
SSPL Science and Society Picture Library
b = bottom; c = center; t = top; l = left; r = right

Jacket
Astronomers examining solar eclipse spectrogram, Roger Ressmeyer/Corbis; Albert Einstein, Bettmann/Corbis; Marie and Pierre Curie in their laboratory, Underwood & Underwood/Corbis; Wegener using a weather balloon to track atmospheric circulation, AKG London.

1 Tony Hallas/SPL; **3** MEPL; **6** Kunsthistorisches Museum, Vienna, Austria/BAL; **7t** Museo Archeologico Nazionale, Naples, Italy/BAL-Roger Viollet; **7b** Araldo de Luca/C; **8** Ted Spiegel/C; **9t** MEPL; **9c** AKG; **10-11** C; **11** BPCC/Aldus Archive; **14** Erich Lessing/AKG; **15** AKG; **16t** British Library, London/BAL; **16b** ARPL; **17** MEPL; **18** AKG; **19** Andromeda Oxford Limited; **20c** AKG; **20b** Erich Lessing/AKG; **21** Gianni Dagli Orti/C; **24** AKG; **25** MEPL; **26** Erich Lessing/AKG; **27c** Guido Sansoni/Biblioteca Nazionale Centrale, Firenze; **27cr** AKG; **27br** MEPL; **28t** British Library, London; **28b** Scala Group SpA; **29** & **30-31** AKG; **32** Scala Group SpA; **33** Guido Sansoni/Biblioteca Nazionale Centrale, Firenze; **34** Archivo Iconografico, S.A./C; **35t** AKG; **35c** Alinari; **38** AKG; **39** MEPL; **40** British Library, London/BAL; **41l** MEPL; **41r** & inset AKG; **42-43** Archivo Iconografico, S.A./C; **44** Erich Lessing/AKG; **44** inset AKG Berlin; **46** Royal College of Physicians; **47t** Erich Lessing/AKG; **47b** British Library, London; **49t** & **49c** Royal College of Physicians; **49b** Mark Fiennes/Royal College of Physicians; **50** Mansell Collection/Time Pix/Rex Features; **51tl** & **51cl** ARPL; **51tr** British Library, London; **51cr** ARPL; **52** Mark Fiennes/Royal College of Physicians; **54** By courtesy of the National Portrait Gallery, London; **55t** The Royal Institution, London/BAL; **55b** Lincolnshire County Council, Usher Gallery, Lincoln/BAL; **56t** Roger Antrobus/C; **56b** Cambridge University Library; **58-59** Alfred Pasieka/SPL; **59** ARPL; **60** Royal Society, London; **60-61** Erich Lessing/AKG; **61** Royal Society, London; **63l** Royal College of Physicians; **63r** ARPL; **64** Royal Society, London; **65** Erich Lessing/AKG; **66-67** Bettmann/C; **67** Private Collection; **70** MEPL; 71 Hammerby House, Uppsala, Sweden/BAL; **72** & **73c** AKG; **73b** David Lees/C; **74-75** Natural History Museum, London/BAL; **76l** AKG; **76r** Carolina Institute, Uppsala, Sweden/BAL; **78** SPL; **79** Bettmann/C; **80** Derby Museum & Art Gallery/BAL; **81t** Bodleian Library, Oxford; **81b** AKG; **82-83** Harlingue-Viollet; **84** Cavendish Laboratory/ University of Cambridge; **85** Erich Lessing/AKG; **86** Bristol City Museum and Art Gallery/BAL; **86-87** SPL; **87** Jean-Loup Charmet; **88** ARPL; **88-89** Mansell Collection/Time Pix/Rex Features; **89** Musée Carnavalet, Paris, France/BAL; **90** Jean-Loup Charmet/SPL; **90-91** AKG; **91** Mansell Collection/Time Pix/Rex Features.